D1042065

THE LONG & THE SHORT OF IT

The Long and the Short of It

The Science of Life Span and Aging

Jonathan Silvertown

THE UNIVERSITY OF CHICAGO PRESS + *Chicago & London*

JONATHAN SILVERTOWN is professor of ecology at the Open University, Milton Keynes, and the author or editor of numerous books, including, most recently, *An Orchard Invisible*.

The University of Chicago Press, Chicago 60637
The University of Chicago Press, Ltd., London

Printed in the United States of America

22 21 20 19 18 17 16 15 14 13 1 2 3 4 5

ISBN-13: 978-0-226-75789-6 (cloth)
ISBN-13: 978-0-226-07210-4 (e-book)
DOI: 10.7208/chicago/9780226072104.001.0001

Library of Congress Cataloging-in-Publication Data

Silvertown, Jonathan W., author.
The long and the short of it: the science of life span and aging / Jonathan Silvertown.
pages; cm
Includes bibliographical references and index.
ISBN 978-0-226-75789-6 (cloth: alk. paper) — ISBN 978-0-226-07210-4 (e-book)
1. Longevity. 2. Life spans (Biology) 3. Aging. I. Title.
QP85.S495 2013
612.6'8—dc23 2013021536

♾ This paper meets the requirements of ANSI/NISO Z39.48-1992 (Permanence of Paper).

For Rissa,
for a lifetime

CONTENTS

1

Death and Immortality

DESTINATION

> Night is the morning's Canvas
> Larceny—legacy—
> Death, but our rapt attention
> To immortality
>
> EMILY DICKINSON[1]

Sooner or later, everyone ponders their mortality. It is the privilege of youth to be oblivious to death, but the fate of old age to contemplate oblivion. Each person searches for answers in his or her own way, but eventually all ask the same questions: How long might I live, and why must I die? What rhyme or reason is there in aging and mortality? Long before science offered reasons, art sought a rhyme that would give meaning to the mysteries of life and death. Such a rhyme is hidden in a priceless and little-known work of medieval art that lies before the high altar of Westminster Abbey in London, England.

Hidden for decades beneath a carpet that used to be rolled back only for the feet of a new monarch, the Great Pavement of Westminster Abbey is a gloriously intricate mosaic floor that depicts a medieval view of the cosmos. It connects the life spans of plants, animals, and people with the life span of the universe and the day of judgment that would herald its end. The story told in the Great Pavement cannot now be read on its damaged surface, but it has been reconstructed by historical and archaeological detective work. An inscription in Latin, which ran around the four sides of the square frame that encloses the pavement, tells us that the mosaic was completed in "this Year of Our Lord 1272," in the reign of King Henry III.[2] The pope contributed to the cost of its construction, and the Italian artisans who laid its dazzling pattern brought with them to dismal London bright stones salvaged from ancient Roman floors: glass tesserae in cobalt blue, turquoise, red and white, and purple porphyry, the livid color of congealed blood. This last is the rarest of the stones in the Great Pavement, found in only one mine in Egypt that closed 500 years before the birth of Jesus.

Within the square frame is a design of four circles that flow into one another, like giant loops formed from a single cord. Around the perimeters of the circles once ran the words:

> If the reader wittingly reflects upon all that is laid down,
> he will discover here the measure of the *primum mobile*:
> the hedge stands for three years,
> add in turn dogs, and horses and men,
> stags and ravens, eagles, huge sea monsters, the world:
> each that follows triples the years of the one before.

Primum mobile refers to the outermost heavenly sphere in the medieval conception of the universe. Thus, according to the inscription, the witting reader will discover in the Great Pavement the measure of the universe or, in other words, how long it will endure. The medieval designers of the Great Pavement knew that different animals and plants have different life spans, and they perceived this variation as part of the grand design of the cosmos itself. The linked circles in the pavement embody the idea that life cycles are yoked together and are linked to the longevity of the universe. Everything was connected by application of the holy number three, culminating in judgment day. The formula that links the life spans in the pavement is three years for a hedge (before it is rejuvenated by cutting), tripled, which gives 3^2 (= 9 years) for the supposed life span of a dog; tripled again for the life span of a horse 3^3 (= 27 years), and so on up to 3 to the power 9, or 19,683 years, for the duration of the *primum mobile*.

Nineteen thousand years must have seemed like a very long time to a medieval cosmologist, but we now know that, looking backward into Earth history, it is scarcely any time at all. The Devonian limestone in the pavement, a rock consisting mainly of fossilized remains of marine creatures, is on the order of 350 million years old, but life has been present on Earth for ten times longer than that (3.5 billion years), and the planet is a billion years older still. The universe is nearly 14 billion years old by current estimates. Although today we are asking the same questions about time that our medieval forebears did, the answers offered by science stretch the imagination to its very limits.

What has science to say about life spans? Why do different species live for such different lengths of time—a dog for maybe 10 years, but a human for 80? Medieval cosmologists believed that there was unity in the diversity of life spans because all belonged to a divinely ordered mathematical series. Does science have its own unified explanation for why longevity varies, or is it just a giant heap of facts, like a pile of mosaic pieces lacking order or design? And what of aging—the dysfunctions that accumulate with age and that terminate even the longest life? Why do we age? Do animals and plants grow decrepit just as we do?

This book is my own mosaic in which I will piece together the answers that modern science offers to these questions. But we will begin in Westminster Abbey because, surprisingly for a medieval church, it has much more to tell us about death and immortality than just the message hidden in the Great Pavement.

Westminster Abbey is where England buries her immortals.

Here death and posterity inhabit the same ground, serving to remind us that great art and scientific understanding transcend mortality. In this place, as much national mausoleum as church, lies Geoffrey Chaucer (d. 1400), author of *The Canterbury Tales*. He is surrounded in Poets' Corner by memorials to William Shakespeare, William Wordsworth, Charles Dickens, Jane Austen, George Eliot, T. S. Eliot, Henry James, and seemingly every other name in the canon of English literature. The walls and floor of this Valhalla are so crowded with illustrious names that there is now an overflow into the stained-glass window above Chaucer's tomb. Oscar Wilde and Alexander Pope are among the names illuminated in the window that lights Chaucer's grave.

But this is an English church, and therefore ironies, rebellion, and even ribaldry run through its solemn fabric like veins in marble. In the seventeenth century, schoolboys from the adjacent Westminster College fought battles in the neglected aisles with the jawbone of King Richard II.[3] Later, young scholars carved their names on tombs and even on the coronation chair, where the graffiti can still be seen. Samuel Pepys, the seventeenth-century diarist, records that the disinterred, mummified body of Queen Catherine of Valois, wife of King Henry V, 232 years dead, was available for display, and that one February day in 1669, "by particular favour . . . I had the upper part of her body in my hands, and I did kiss her mouth, reflecting upon it that I did kiss a Queen."[4]

Signs of such sacrilege horrified later visitors. Washington Irving, visiting from New York at the beginning of the nineteenth century, wrote:

> What, thought I, is this vast assemblage of sepulchres but a treasury of humiliation; a huge pile of reiterated homilies on the emptiness of renown and the certainty of oblivion! It is, indeed, the empire of death; his great shadowy palace where he sits in state mocking at the relics of human glory and spreading dust and forgetfulness on the

monuments of princes. How idle a boast, after all, is the immortality of a name![5]

When one is surrounded in the abbey by a thousand forgotten names, it is tempting to agree. How can any human life span, ending as it must in aging and infirmity, measure up against the eternity of death? In the South Aisle, around the corner from more famous poets, is the memorial to William Congreve (1670–1729), a poet and playwright whose pallbearers included the then prime minister, but who is now scarcely remembered.[6] Congreve's lover, Henrietta, Duchess of Marlborough, spent part of his legacy to her on a mechanical statue of Congreve, carved from ivory and driven by clockwork. The duchess talked daily at table to her wind-up lover, as though he were still alive, lending his memory, at least for her, a temporary reprieve from death.

The abbey is also the church where the kings and queens of England are habitually crowned. The pinnacle of pomp was reached here in 1902, at the coronation of Edward VII, when the British Empire was at its zenith.[7] The king of England and of a quarter of the planet, the emperor of India, was warned by his doctors before the event that he could die during the ceremony if he didn't delay it to be treated for acute appendicitis. Reluctantly, the sovereign bowed to his own mortality, but was still weak when the coronation eventually took place. Rank and title are no protection from the infirmities of old age. The 80-year-old archbishop who performed the ceremony was in even worse condition than the king. Half-blind and with trembling hands, he had difficulty reading the service and had scarcely the strength to lift the crown onto the head of the new monarch. The king and three bishops had to help him to his feet after he knelt before the throne. The archbishop died within months. King Edward VII died only eight years later at the age of 68.

How is King Edward VII remembered today? The coins issued during his reign, surely durable and numerous enough to perpetuate Edward's name for centuries, are long out of circulation. British schoolchildren no longer memorize the names and dates of the monarchs that their great-grandparents learned by rote. In 1902, however, a vegetable grower honored King Edward by naming a new variety of potato after him. So, ironically, in England now, King Edward is a spud. Potatoes live longer than kings. Each potato tuber is genetically identical to the plant that made it, and since each crop is grown from tubers saved from a previous one, the original King Edward potato is still alive, multiplying with every season. The Idaho potato is an even older variety, used to make the fries served in McDonald's restaurants. These potatoes will outlive us all, especially if we eat too many of them. We will discover later why plants break all the records for extreme longevity and how diet influences life span in animals, including ourselves.

Notwithstanding the poignant examples of the fickleness of fame, Washington Irving was wrong. Some names, including his own, are remembered. Will Shakespeare ever be forgotten? Who can fail to recognize the handle of George Frideric on the composer's tomb in Poets' Corner while his sublime music still roars? Creators of immortal works live on, even if Woody Allen did once quip, "I don't want to achieve immortality through my work. I want to achieve it by not dying." A remark that would probably not have amused Sir Isaac Newton, who was better known for gravity than levity. It is said that he laughed only once in his entire life, when someone asked him what use he saw in Euclid's *Elements*.[8] Newton's marble memorial in the abbey is so elaborate it looks like a shrine, as well it might for this luminary of science. Alexander Pope famously wrote in his eulogy of Newton, "Nature and Nature's laws lay hid in night: God said, 'Let Newton be!' and all was light."

A few paces from Newton's shrine is the stark burial place of Charles Darwin, covered by a simple floor slab of white marble inscribed only with his name and dates. By the time of Darwin's death, the Anglican Church was largely reconciled to the theory of evolution, and it had been added to the list of nature's laws ordained by God. As for Darwin himself, although he had trained for the clergy as a young man, he died an agnostic. Darwin's faith foundered on two questions that still trouble religion today. Why does God permit evil? And where is the material evidence of God's existence? Charles Darwin was a man of great sensibility and kindness, devoted to his family, fiercely opposed to slavery, and considerate of others. When his beloved daughter Annie died of tuberculosis at the age of 10,[9] he could not imagine how God, if he existed, could tolerate the suffering of an innocent child. Charles's wife, Emma, found consolation for Annie's death in religion, but Darwin found only doubt. Today the scientific puzzle is why evolution permits aging and death. Why me, oh Lord, but not the ageless Idaho potato?

Side by side with Darwin's tomb in Westminster Abbey, so close that the tombstones touch, is the grave of Sir John Herschel, astronomer and mathematician. Long before Darwin published his book *The Origin of Species*, Herschel had pondered what he called the "mystery of mysteries, the replacement of extinct species by others" and speculated that "the origination of fresh species, could it ever become under our cognizance, would be found to be a natural in contradistinction to a miraculous process." When he came to write *The Origin of Species*, Darwin referred to Herschel's comment on the "mystery of mysteries" in the introductory chapter. The title Darwin chose for his book may also have been inspired by Herschel's phrase "the origination of fresh species." Darwin's great achievement was to discover how new species can arise naturally, without miraculous creation. He discovered how evolution happens.

Darwin called the mechanism that drives evolution natural selection. Let individuals vary, he said, and those that are better equipped for survival in the struggle for existence that characterizes everyday life will leave more offspring than their lesser companions. Now imagine that the variation on which this winnowing of nature operates is inherited and is passed from parents to offspring. Then, those characteristics that lead to greater reproductive success will be naturally selected and will increase in each generation. Over many generations, natural selection will produce change, and given sufficient time, as Darwin wrote in the closing lines of *The Origin of Species*, "endless forms most beautiful and most wonderful have been, and are being evolved."[10]

Westminster Abbey is a testament to the struggle for existence, for in this building we see how great a force is mortality. You cannot enter the abbey, first constructed more than a thousand years ago, without being reminded of the brevity of human life compared with the immensity of time. Until recently, disease was a great harvester of young life and talent, so much so that if those commemorated in Poets' Corner were to be resurrected in that spot, a significant part of it would become a tuberculosis ward.[11] John Keats (d. 1821) died at age 26 of the disease. It also killed at least two of the three Brontë sisters, their wayward brother Branwell, Elizabeth Barrett Browning (d. 1861), and D. H. Lawrence (d. 1930). Alexander Pope (d. 1744) suffered stunted growth and lifelong illness associated with TB. Other literary sufferers included Robert Burns (d. 1796), Henry David Thoreau (d. 1862), and Washington Irving (d. 1859). The bacillus that causes TB has left its evolutionary mark on the human genome.[12] Natural selection has increased the frequency of resistance genes in human populations that have been most exposed to the disease. In fact, the human genome is peppered with genes that have a function in protecting us from disease, all of them the product of natural selection driven by the epidemics of the past.[13]

Death in childbirth was once also very common and was no respecter of rank.[14] King Henry VIII's mother and two of his six wives died this way. Scarlet fever, a bacterial disease, carried off children of the privileged as well as those from more humble families. In Louisa May Alcott's famous novel *Little Women*, set at the time of the American Civil War, 13-year-old Beth March catches scarlet fever while helping the poor and eventually dies of the disease. Mortality was so ever-present in her world that even Beth's six dolls were all invalids. Vaccination, antibiotics, and good sanitation and health care have freed inhabitants of the developed world from the everyday fear of maternal and child death, but tuberculosis is still the biggest cause of preventable deaths in the developing world.

Science and public health have won important battles against infection, but not the war. Bacteria have very short generation times, which gives them the ability to multiply and to evolve at enormous rates. For example, the bacterium *Helicobacter pylori*, which usually lives harmlessly in human stomachs but can cause stomach ulcers and even cancer, is usually acquired during childhood and, if not treated, evolves genetically distinct strains within the human body during a lifetime of infection.[15] Half the human population carry this bacterium, and if you and I are both infected, mine will certainly be distinct from yours. The ability of short-lived pathogens to evolve rapidly has led to the appearance of genes for antibiotic resistance in *H. pylori*, in the tuberculosis bacterium, and in many others. These genes spread because they enable the bacteria that possess them to survive our attempts to poison them, but worse still, they can be transferred between unrelated bacteria, so antibiotic resistance can spread very rapidly and form combinations that elicit those words that you never want to hear your doctor utter: multiple drug resistance.

Other animals have different species of *Helicobacter*, but strangely enough, the animals with the *Helicobacter* bacteria that are genetically most like our *H. pylori* are not, as you might expect, our closest primate relatives, such as chimpanzees or monkeys, but big cats like cheetahs, lions, and tigers. It is estimated that the ancestor of this *Helicobacter* jumped from humans to big cats about 200,000 years ago, when our species was still at home on the range in Africa.[16] Back then, fear of big cats would certainly have given our ancestors stomach ulcers. It appears that, thanks to *H. pylori*, we returned the compliment.

In disease-causing bacteria, we begin to see the evolutionary significance of life span. In fact, it is not short life span as such, but short generation time that confers such an enormous advantage on bacteria. Life span is the average time between birth and death. Generation time is the time between being born and having offspring. Bacteria reproduce by division, so for them life span and generation time are one and the same and can be as short as 30 minutes. Human generation time is on the order of 20–25 years, whereas life span is 70–80 years.

Short generation time spins the wheels of evolution faster and makes rapid evolution possible, which is one reason why bacteria can adapt so quickly to new challenges such as antibiotics. But, even ignoring this capacity for adaptation, short generation time is numerically advantageous. The winners of the evolutionary game are those who leave the most descendants, and because having a short generation time speeds the rate of increase, it confers a massive advantage. While longer-lived organisms are going through their difficult teenage years, short-lived ones are having babies and their babies are having babies. But herein lies a puzzle. If short generation time is so advantageous, why isn't it universal?

The mosaic I have created in this book is framed by a series of linked puzzles, of which this conundrum of longevity is just the first. These puzzles require a whole heap of curious facts and ingenious arguments to solve them. Even if you are interested only in your own species, you will discover in chapter 2 that the answer to why we aren't all as short-lived as microbes will be found by comparing many species, because each is like an experiment in evolution with something potentially new to tell us. Next, in chapter 3, we inquire, "What is aging, and how long could we live if it were possible to abolish it?" Chapter 4 investigates the influence of inheritance on longevity, revealing the startling fact that tweaking particular genes that are shared by all animals can dramatically lengthen life.

The Greek philosopher Aristotle (384–322 BC), who has also been called the first biologist for his direct observations of the natural world, wrote very perceptively about longevity. He observed that plants were the longest-lived organisms because they can "continually renew themselves and therefore last a long time." In chapter 5 the puzzle is how plants, from potatoes to giant sequoias, manage something that very few animals achieve. With the knowledge that genetic modification can lengthen life and that some plants appear virtually immortal, Chapter 6 tackles the biggest puzzle, which is why death exists at all, or more exactly, "Why does natural selection, which favors organisms that survive and reproduce, ever permit aging and death?" In chapter 7 we probe the big question that suicidal species such as Pacific salmon raise: "Can death ever be adaptive?" Finally, in the last two chapters, we arrive at a thicket of puzzles about how aging occurs at the molecular level. So many things go wrong as bodies age that even picking the right questions to ask becomes difficult. Yet there is meaning in this madness. This is the outline of the loose pieces in my mosaic. If you wish to see how they fit together and the grand pattern that they make, let me roll back the carpet for Your Majesty, and then follow me.

2

An Hourglass on the Run

LIFE SPAN

And what is Life? An hourglass on the run
A mist retreating from the morning sun
A busy bustling still repeated dream
Its length? A moment's pause, a moment's thought
And happiness? A bubble on the stream
That in the act of seizing shrinks to nought

JOHN CLARE, "WHAT IS LIFE?"[1]

When the rustic poet and naturalist John Clare (1793–1864) wrote these words, the life of an agricultural laborer such as he was indeed, in the words of Thomas Hobbes, "poor, nasty, brutish and short." Yet even such a life is a pinnacle of longevity by comparison with that which falls to the lot of most organisms. From the point of view of evolution, there is not a lot to be said for long life. Natural selection favors inherited characteristics that assist reproduction, so a gene for short life and early reproduction will spread like wildfire as its carrier's offspring multiply and their offspring, in turn, multiply down the generations. By comparison, an organism carrying a gene for later maturity and longer life will plod slowly toward posterity and fast become history. It is just arithmetic. Imagine two banks that pay you compound interest on your savings. Which will earn you more, one that pays 5 percent a month or one that pays 5 percent a year? Monthly compound interest at 5 percent will turn $100 into nearly $180 in a year, a return forty times better than the $5 you will get from the tardy bank. That's exactly the kind of advantage that short life and early reproduction confers on organisms. And by the way, if you find a bank paying even 2 percent a month, be sure to let me know.

The puzzle of longevity, then, is not why we die so soon, but rather why we live so long. There is, of course, a solution, but it took evolution more than 2.7 billion years to discover it, so we need to begin near life's own beginning. The first organisms to evolve were simple and bacteria-like, and for a great deal of life's history on Earth, that's all there were. The first signs of life appear in the fossil record from about 3.5 billion years ago, and microbes were Earth's only inhabitants for the next 2.7 billion years. The world's shortest poem, entitled "Lines on the Antiquity of Microbes," pithily celebrates this fact:

Adam
Had 'em.

These were solitary cells or, at their most complicated, chains or sheets of identical ones. Everything we commonly think of as life today—all the organisms that as individuals are big enough to see with the naked eye—evolved in the last 800 million years.

So part of the solution to the puzzle of longevity is that, for a long, long time, there simply was no puzzle. For the majority of Earth history, nearly every organism was single-celled[2] and was, at least potentially, short-lived and able to multiply fast. Even today microbes are numerically dominant. The cells in your body are outnumbered at least ten to one by the bacterial and fungal cells that call your body home.[3] The American poet Walt Whitman wrote in his "Song of Myself" (1855): "I am large, I contain multitudes."[4] He cannot have known how right he was.

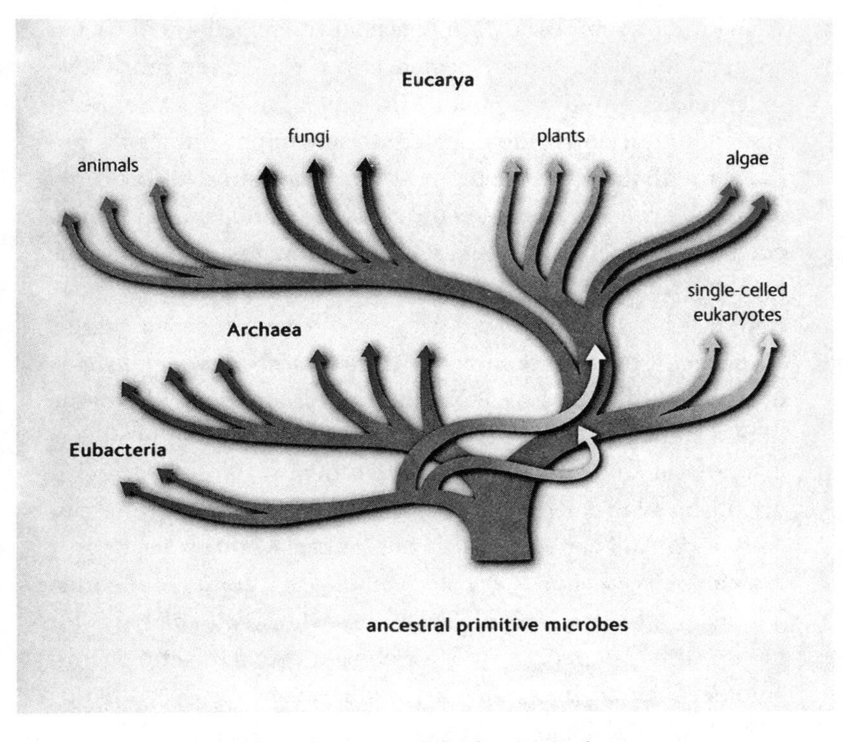

FIGURE 1. The tree of life, showing its three principal branches, the Archaea, the Eubacteria, and the Eucarya. (From 99% Ape.)

Of the three major branches of the tree of life, two, the Eubacteria and the Archaea, are composed only of microbes. Even on the third branch, called the eukaryotes, from which the human species is a tiny, recent offshoot, there are many single-celled organisms (fig. 1). Microbes are astonishingly diverse in their genes and in their biochemical talents. In the time that they had the planet to themselves, they managed to evolve every imaginable means of making a living, including capturing the sun's energy through photosynthesis, deriving energy from chemical reactions involving sulfur in the sunless, abyssal depths of the ocean, surviving in water hot enough to poach an egg in Yellowstone's hot springs, and eking out an existence in rocks buried nearly two miles underground in a South African gold mine.[5] The appearance of multicellular life opened up new opportunities for microbes, both in and on the curiously tardy upstarts. No cow could thrive on grass, no termite could conquer wood, and no human could survive without the microbes[6] that process our food for us in our guts.

There is a limit to how big a single-celled organism can be. The biggest bacterium known is the sulfur pearl, which lives in ocean mud off the coast of Namibia and is about the size of the period at the end of this sentence.[7] When multicellular life finally evolved from the primordial flea circus of microbial talent, bigger organisms with longer lives became possible, but these organisms are all still coalitions of tiny cells. Liza Minnelli sang, "Life is a cabaret, old chum," but it's not. Life is a condo, old chum. Though, to be fair, I expect Fred Ebb, who wrote the lyrics for the musical *Cabaret*, would have choked trying to get the word "multicellularity" into a song that made any money.

Not only did life itself begin with a single cell, but each of our individual lives begins that way too—with the single cell that is a fertilized egg. That cell divides, and the embryo grows and develops in a highly coordinated way that is so faithful to its inherited plan that it produces strong family resemblance between parents and offspring. That multicellular organisms are condominiums of cooperating cells is of vital significance to their longevity. The upside of being multicellular is that an organism can repair itself by using new cells to replace damaged, worn-out, or infected ones. Specialized immune cells fight infection by identifying, engulfing, and destroying pathogens. Thus a multicellular organism has a repair unit, a cellular defense force, and a health service, all of which help to prolong life.

A potential downside of being multicellular is that for growth and repair to take place, some cells must retain their inherent capacity to divide, but if these so-called stem cells are allowed to proliferate in an uncontrolled manner, the result is cancer. Letting cell division rip is a hazard that shortens life. About a quarter of Americans die of cancer. Multicellular organisms have numerous mechanisms for controlling cell proliferation and preventing cancer, but all of these mechanisms ultimately depend on the action of genes that act as control points, or brakes on runaway cell division.

It is as if every multicellular organism is an automobile parked on a steeply graded street that ends with a fatal plunge into San Francisco Bay. Multiple precautions guard against disaster. There is a local ordinance in San Francisco stating that if you are parked on such a street, you must turn your wheels into the curb; then there is the hand brake, and putting the car in "park," which uses the transmission to lock the wheels. Cells have more devices that stop them careering out of control than cars do, but since there are many billions of cells, each the product of cell division, the cancer hazard for an individual is more like the chance of one runaway occurring in a thousand San Franciscos, all full of autos poised on vertiginous streets. No wonder, then, that most of us will have tumors in our bodies when we die, even if they are not the direct cause of death. We are up against the overpowering math of cell division.

Just how powerful the multiplication of unchecked cancer cell division can be is illustrated by a line of cells known as HeLa, after a cervical cancer patient named Henrietta Lacks, from whose cancerous tissue they originated in the early 1950s. Until the discovery of HeLa, scientists had been unable to keep human cells alive and dividing in laboratory culture for any length of time. There seemed to be an inherent limit to the number of times cells taken from a multicellular animal would divide before they gave out, but the cells from Henrietta Lacks's papilloma behaved like they had not read the textbook—when provided with the right conditions in the lab, they divided and divided and divided.

The HeLa cell line quickly became an important tool in biology and medicine. Just a year after Henrietta's death in 1951, her cells were used in testing the new polio vaccine that eventually saved millions of lives. Within a few years, the lab supplying HeLa cells was shipping 20,000 tubes, containing a total of maybe 6 trillion cells, a week. The HeLa cell line became so ubiquitous, and grew so easily, that it began to contaminate lab cultures of other cells, behaving more like a microbe than a human cell line. Indeed, more than one evolutionary biologist has proposed that HeLa be recognized as a new species, since it has an autonomous existence.[8] HeLa cells even have their own best-selling biography, *The Immortal Life of Henrietta Lacks* by Rebecca Skloot.[9]

HeLa is not alone among tumor cells in throwing off the trammels of a multicellular existence to roam like a vagabond, freed from the rules that govern life in a condominium. There is a venereal disease in dogs caused by infectious cells that form tumor-like growths on the genitals of affected animals. The disease is worldwide, affects all dog breeds, and has even been found in foxes, but all infections seem to be caused by the same cell line, which had a single point of origin.[10] Fortunately, these canine venereal tumors regress within a few months, presumably because they are attacked by the host's immune system, which rejects the foreign tissue, just as organ grafts are rejected in humans unless drugs are taken to suppress the immune system.

The immune system is one of the chief defenses that multicellular animals have against infection, but immune cells require a genetic signature to be able to distinguish friends—other inhabitants of the condominium—from foes. In genetically variable populations such as our own, each individual has his or her own genetic signature (although there are many similarities among close relatives), and the system works well. In highly inbred populations, however, genetic variation is much reduced, and this can open the door to vagabond tumor cells. In 1996 a new disease suddenly afflicted the Tasmanian devil, a marsupial carnivore whose remnant populations are found only on the island of Tasmania. The disease causes tumors on the faces of afflicted animals and is invariably fatal. Researchers made the startling discovery that the cells taken from tumors on different animals were uncannily alike, suggesting that the tumors did not originate independently from a different rogue cell in each individual, as most cancers do, but were being transmitted from animal to animal by muzzle contact during fighting.[11] Populations of rare island species such as the Tasmanian devil are often highly inbred, which is probably why this species, now listed as endangered because of the disease,[12] is susceptible to vagabond tumor cells. The Tasmanian devil may have entered what conservation biologists call an extinction vortex, where small population size leads to inbreeding and inbreeding exposes individuals to disease, further reducing population size to a point where it is so small that chance alone can finish it off.

Cancer in all its forms is a brutal reminder that long life is a precarious achievement that has to be defended against the unfettered force of rapid cell division. Cancer risk is the price of multicellularity in animals and the longer life that it makes possible. What makes cells go rogue? The source of the problem lies mainly in spontaneous changes in the genetic code, written in DNA, that controls how a gene functions. These changes are called somatic mutations. Each time a cell divides, the DNA is replicated, so that each new cell contains its own copy of the code. Copying errors are rare, but they do happen, and even a rare event can become inevitable if you give it enough chances. During a single week, all the cells in the surface lining of your gut will have been replaced by cell division—twice.[13] One somatic mutation is not sufficient to unleash a cell from the controls on cell division, but it can fray a brake cable and start those cell's descendants on a pathway toward eventual cancer. By the age of 60, some of the stem cells in your gut that supply replacement cells may have divided 3,000 times. Multiply this figure by the tens of millions of stem cells in the gut and, if mutations are not controlled in some way, some people will have accumulated hundreds of mutations by the time they are 60. How, then, does anyone make it past 60?

Cancer is a hazard of multicellularity in animals, but it does not determine how long different species live, and this is another puzzle. When death rates from cancer are compared between animal species, they do not vary as much as one might expect. For example, cancer kills about 20 percent of dogs, 18 percent of beluga whales, and as we have already observed, 25 percent of humans in the United States. This trivial variation among species is remarkable because cancer rates seem to bear no relationship either to how long the different species live—from about 10 years for dogs to 40 years for beluga whales to about 80 years for humans—or to how big they are (belugas can weigh up to a ton and a half). Cancer rates ought to increase with size and longevity because both expose an animal to greater risk of the somatic mutations that trigger uncontrolled cell division. Greater size increases this risk because bigger animals have more cells than smaller ones. Longer life increases the risk because it requires more cell replacement through cell division. Both greater age and greater size should therefore incur a greater risk of at least one cell turning cancerous, with potentially fatal consequences.

Let's try a simple back-of-the-envelope calculation to check out this argument. The incidence of colorectal cancer in 90-year-old humans recorded by the American Cancer Society is 5.3 percent.[14] Mice have roughly a thousand times fewer cells than humans, which would mean that even if they lived to 90 (at least thirty times their actual longevity), their chance of dying from colorectal cancer would be a thousand times less than for a 90-year-old human, or an infinitesimal 0.0053 percent. A blue whale, on the other hand, with a thousand times the body weight and number of cells of a human, ought to have a thousandfold higher rate of colorectal cancer by age 90—so high a rate, in fact, that by 80 years of age all blue whales would have succumbed. Yet blue whales, the largest animals on the planet, are not the floating vessels riddled with piratical tumors that these calculations suggest they ought to be. And as for mice, the eminent cancer epidemiologist Richard Peto observed, in an article called "Cancer and Ageing in Mice and Men," published back in 1975, that "most species suffer some cancer of old age, whether old age occurs at 80 weeks or 80 years."[15] This observation is now called "Peto's paradox."

Peto's paradox has the clear implication that somehow longer-lived species are better protected from cancer than shorter-lived ones and, likewise, bigger species are better protected than smaller ones.[16] If cancer rates increased with size and age as species evolved, no animal could live longer than a mouse, and the bowhead whale could certainly not attain the record for vertebrate longevity of 200 years.[17] There is only one way of explaining Peto's paradox: evolution can modify susceptibility to cancer. This conclusion is now supported by evidence that the genes protecting us from cancer are also associated with long life.[18] Peto's paradox is more than just a curiosity of comparative biology; it is a pointer to where to look for effective countermeasures against cancer in the animal world. Maybe the message hidden in the Great Pavement of Westminster Abbey was correct and the secret of longevity really is in huge sea monsters.

Now that we are in the realm of multicellular creatures, let's explore how life span varies among species and try to find out why it does so. Do you need to be big to live long? The connection between size and life span in the animal world was obvious to Aristotle over 2,000 years ago, but is size a cause or a consequence—perhaps an incidental one—of long life? Large size could be a direct cause of longevity if being big protects an animal from predators that would like to call it lunch, or if it helps an animal to survive cold winters. On the other hand, it takes time to grow large, and if large size offers other advantages that have nothing to do with survival, such as greater reproductive success, they would provide an incidental reason why longevity and size are correlated.

Of course it is also possible that both direct and indirect causes link size and longevity. This is probably what happens in bivalves (clams, mussels, and oysters), which continue to grow throughout life. As the shell grows thicker and bigger, the animal inside becomes better and better protected, potentially leading to a very long life. Growth rings on the shell indicate age, just like the growth rings in a tree trunk, and from these rings it has recently been discovered that bivalves are some of the longest-lived animals on the planet, rivaling and even beating bowhead whales and giant tortoises.[19] A geoduck clam, found in the coastal waters of Washington State and British Columbia, clocked up 169 years, and the European freshwater pearl mussel beat that at 190 years, but the granddaddy of them all is a 405-year-old specimen of the ocean quahog from the coastal waters of Iceland.

How Aristotle came to the conclusion that large animals live longer than small ones we do not know. He conducted zoology fieldwork in a lagoon on the Greek island of Lesbos, where he dissected various marine animals. Without access to a microscope or even a lens, however, he could probably not determine the ages of fish in the way that a modern zoologist would by using the growth rings in their scales. Perhaps he observed that small fish species breed at an earlier age than big ones. The extreme in this trend, discovered only recently on the Great Barrier Reef in Australia, is the stout infantfish, which breeds at a quarter of an inch long and dies after only two months, when other fish this size are still fry in the cradle.[20] More likely, Aristotle relied on his knowledge of domestic animals such as dogs, goats, and horses.

Dependable data on how long different wild species live are very hard to obtain, and only recently has it been possible to accurately compare the longevities of many species. Some data come from records of zoo animals, but these can be biased upward by protection from the hazards that the species would normally face in the wild, or downward by the effect of poor conditions in captivity. Furthermore, the number of individuals in most of these zoo samples is small, making estimates less accurate. The best estimates come from field studies that catch, mark, release, and later recapture lots of animals over long periods. Good data on vertebrates of all kinds have been collected this way.

Comparisons involving hundreds of species of mammals, birds, and reptiles show that Aristotle was right about the correlation between size and longevity, but only as a broad generalization. Among mammals, for example, bigger species do indeed live longer than smaller ones on average, but there are plenty of notable deviations from the general trend.[21] For their size, marsupials (opossums, kangaroos, and relatives) have shorter lives than placental mammals, while at the other extreme, primates, the mammal order to which we belong, live longer than mammals of similar size in other orders. Bats also live much longer than nonflying mammals such as rodents. The common pipistrelle bat weighs less than a fifth of an ounce as an adult, but has been known to live for 16 years in the wild, while the house mouse weighs four times as much and lives only one-fourth as long (4 years), at most.[22]

There are exceptions among rodents too. The naked mole-rat is an extraordinary creature that lives in underground nests in family groups dominated by a queen who is cared for by nonreproductive workers, just like bees in a hive. Naked mole-rats are only the size of a large mouse, but they can live for up to 28 years.[23] This is an astonishing age for so small a rodent. Imagine buying your kid a small rodent such as a hamster as a pet, say for her fifth birthday, and her having to care for it until she is into her thirties. The trauma of the experience might cost you grandchildren. The longevity of naked mole-rats is especially remarkable because rodents as a group are not long-lived. The biggest rodent of all, the capybara, reaches 110 pounds in weight but lives to only about 10 years in the wild.

Birds, like bats, have unusually long lives for their size, living about 50 percent longer than an average mammal of the same size.[24] Perhaps vertebrates that can fly, whether birds or bats, live longer because flight helps them escape predation. There is, of course, also great variation among birds in how long they live, and much of this variation, though by no means all of it, is correlated with size. The flamingos and their relatives are the longest-lived birds, with the parrots not far behind, and the seabird group to which petrels and albatrosses belong running a close third.[25] Not surprisingly, because of their small size, perching birds like thrushes and sparrows, belonging to the order known as the passerines, are short-lived, but within this group the crows are an exception, living 17 years or more on average. Crows have been known to fashion tools to reach food,[26] and for their intelligence, social organization and longevity they might be called the primates among passerines.

Gerontologists, for decades given to self-absorbed gazing at the shriveling navel of their own species, are now increasingly interested in what makes naked mole-rats, ocean quahogs, and other exceptional inhabitants of Methuselah's menagerie tick, and tick for so long.[27] As primates, we live much longer than the average mammal, and even for a primate, we are generously endowed with years, but just how long can a human live? Arriving at an answer to this question is not as easy as you might think. At least bivalves do not tell tall tales.

If his memorial is to be believed, the oldest person buried in Westminster Abbey is Thomas Parr, or "Old Parr" as he was known because he was said to be 152 when he died in 1635.[28] In the seventeenth century, as now, there were those who were all too happy to exploit another's celebrity for their own ends. Old Parr shot to stardom and was extinguished all within a single year. In 1635, when he was blind and all but toothless, Parr came to the attention of Thomas Howard, fourteenth earl of Arundel. The earl was a collector of antiquities of all kinds, and he had Parr carried from his native county of Shropshire in a litter that was met along the way by admiring crowds as it progressed by stages to London, where Parr was exhibited and presented to the king.

Seizing the opportunity offered by the arrival of a celebrity, a poet named John Taylor published a biographical pamphlet in verse, which he titled *The Old, Old, Very Old Man*,[29] not leaving his readers in any doubt about its subject matter. Taylor did not fail to mention the details his readers were sure to appreciate, such as Old Parr's adultery at the age of 105 with "Fair Katherine Milton,"

> Whose fervent feature did inflame so far
> the Ardent fervour of old Thomas Parr.

Even 47 years later when he appeared in London, Taylor tells us, Old Parr had most of his faculties:

> He will speak heartily, laugh, and be merry;
> Drink Ale, and now and then a cup of Sherry;
> Loves company, and Understanding talk,
> And (on both sides held up) will sometimes walk.
> . . .
> And thus (as my dull weak Invention can)
> I have Anatomiz'd this poor Old Man.

It could not end well, and the rich living of London, or the pollution of the city, killed Old Parr before the year was out. Having been anatomized in life by John Taylor, he was then anatomized in death by the most famous surgeon of the age, John Harvey, discoverer of the circulation of the blood.

Who can resist a good story? John Taylor's biography of Thomas Parr was reprinted many times, and its hero entered folklore. Two centuries later the story was further embroidered in a new pamphlet entitled *The Extraordinary Life and Times of Thomas Parr*. This pamphlet claimed that Old Parr's last will and testament had just been discovered and that it contained the secret of his longevity: a herbal concoction that could be purchased in the form of "Parr's Life Pills for Health, Strength and Beauty." These pills were still being advertised in 1906.

Seventeenth-century philosophers were intensely interested in discovering how to prolong life. The French philosopher René Descartes (1597–1650) became suddenly preoccupied with life extension at the age of 41 when he noticed that his hair had started to turn gray. The human mind has enormous difficulty comprehending its own extinction, whatever science and common sense have to say about the matter. Despite convincing himself that it should be possible to live as long as the patriarchs of the Old Testament, and giving others the impression that he fully intended to do so, Descartes died of pneumonia at the age of only 53. An unsympathetic newspaper commented at the time that "a fool has died who claimed to be able to live as long as he liked."

Unfortunately, like those of René Descartes, the researches of other seventeenth-century natural philosophers interested in longevity did not seem to prolong their own lives either. Francis Bacon (1561–1626), to whom we owe the observation "knowledge is power," wrote *History of Life and Death—the Prolonging of Life*, in which he catalogued the names of people who had lived to a great age and recommendations for how to emulate them.[30] He himself died at the age of 65, probably another victim of pneumonia, after catching a chill during an impromptu experiment to see whether stuffing a chicken carcass with snow would preserve it from decay. Incidentally, he was right, of course, and almost exactly 200 years later Clarence Birdseye (1886–1956) made his fortune with a patent for fast-freezing food.

Robert Hooke (1635–1703), a polymath who coined the term "cell" for the tiny chambers that he drew in his pioneering book of observations made with a microscope, had an ingenious, if flawed, explanation for why no one seemed to reach the extreme ages recorded in the Old Testament anymore.[31] Hooke suggested that the ages reached by Adam (930 years), Methuselah (969 years), or even that youngster Abraham (175 years) were measured in shorter years than now because the traverse of Earth around the sun has been slowed by friction since biblical times, lengthening the year. Thus the patriarchs only *apparently* enjoyed longer lives because those lives were measured in shorter years.

There is no documented case of any human living to anything like 152 years of age, so we must take Old Parr's claim with a cellar of salt, but he has the nobler and surer distinction of being the only farm laborer ever to be buried among the rich and famous in Westminster Abbey. The life of farm laborers 400 years ago would normally have been grindingly hard and brutally short. Even today, people in manual occupations have shorter lives than those of higher social status. All the more strange, then, that there have been many claims that the oldest people in the world are to be found in poor, rural communities living lives of hardship in remote places. It is as if the paradise lost in the garden of Eden is there for the seeking in some mountainous Shangri-La where age is defeated by honest toil and modest living.

Among the legion of his books for children, the immortal Dr. Seuss wrote a single work for grown-ups called *You Are Only Old Once*, in which he compared the sorry condition of the aged in normal places with their state in his own imaginary Shangri-La:

> In those green-pastured mountains of Forta-fe-Zee
> Everybody feels fine
> at a hundred and three
> 'cause the air that they breathe
> is potassium-free
> and because they chew nuts
> from the Tutt-a-Tutt Tree

This gives strength to their teeth
it gives length to their hair
and they live without doctors
with nary a care[32]

Dr. Seuss's Forta-fe-Zee might have been inspired by the mountain village of Vilcabamba in Ecuador, a once-celebrated Shangri-La in the Andes where supercentenarians (people aged 110 and over) flourished. Grace Halsell, author of the book *Los Viejos: Secrets of Long Life from the Sacred Valley*, says that she went to Vilcabamba and asked the inhabitants to "take me in."[33] They generously obliged in more senses than one. Manuel Ramon, who said he was 110, climbed mountains like a goat; Micaela Quezada boasted of her virginity at 104; Gabriel Erazo claimed to be as horny at 132 as he had been at 20. For a while claims like these made Vilcabamba a mecca for medics and others studying aging. In the end, however, the documentary records that were at first thought to support the claims of extreme old age in Vilcabamba were found wanting.[34] None of the old people had even reached 100—their average age was in fact 86. A study of life expectancy that compared the population of Vilcabamba with that of a nearby town discovered no difference between the two and revealed an average value that was 15 to 30 percent lower than in the United States.[35]

All around the world, in remote parts of Pakistan, China, and Azerbaijan, one supposed Shangri-La after another has proved to be as fictional as the original Shangri-La itself, built on exaggeration and credulity.[36] In Greece, as recently as 2010, the government discovered that 300 of the 500 recipients of pensions paid to supercentenarians were in fact dead. In the United States, fewer than 25 percent of the people recorded as being more than 110 when they died could be proved to have been genuine supercentenarians. As the editor of the *Guinness Book of World Records*, who was often called on to adjudicate claims of extreme age, has written, "No single subject is more obscured by vanity, deceit, falsehood and deliberate fraud than the extremes of human longevity."

So much for the myths, but what of the reality? At the time of this writing, the oldest person whose age has been substantiated was Frenchwoman Jeanne Calment, who was 122 years, 5 months, and 2 weeks old when she died in 1997.[37] She was born and lived in Arles, in the south of France, where Vincent van Gogh, whom she met at the age of 13, painted some of his most famous work. The oldest man of verified age to have lived is Christian Mortensen, a Danish American who died at the age of 115 in 1998. He is one of the only two males to make it into the top 20 list of longest-lived people.

Most of these oldest of the old became progressively frail after age 105, but not so Jeanne Calment. At 90 she made a contract with a lawyer who agreed to pay her an annual fee for the option to buy her house when she died. He paid up for 30 years before predeceasing her when he was only 77. At age 110 Jeanne moved to a nursing home, not because of illness, but because she nearly burned down her house. One very cold January day the water in the boiler froze solid, so she climbed onto a table and tried to thaw it out with a lighted candle, setting fire to the insulation. Though Jeanne was reluctant to move, she must have seen the funny side of the incident, because her recipe for a long life was "Always keep your sense of humor. That's what I attribute my long life to. I think I'll die laughing." She loved the celebrity that her age brought her and liked to joke, "I've never had more than one wrinkle, and I'm sitting on it."

There are remarkably few similarities among the healthful old, but they do seem to include a lively sense of humor. When Dan Buettner, a journalist working for *National Geographic*, tried interviewing a 91-year-old shepherd named Sebastian for an article about the secrets of healthy old age in Sardinia, he was given a hard time: "I approached him and opened the conversation by asking him his age. 'Sixteen,' he responded, wearing a prankster's smile. Thinking we might break the ice by buying him a round, we asked him if he drank. 'No, my doctor told me not to drink—milk that is.' "[38] I suspect that the oldsters of Vilcabamba had a good laugh at their interrogators' expense too.

Well might the oldsters laugh, for they place us humans in the select company of the other exceptional species that live longer than their close relatives, even the bigger ones. In the museum of the long-lived, we belong right there behind the glass with bats and flamingos, the naked mole-rat, the ocean quahog, bowhead whales, and the latest recruit to Methuselah's menagerie: the olm, or human fish. This animal, neither primate nor fish, is in fact a tiny, three-quarter-ounce, blind salamander found in caves in Eastern Europe. It is reckoned to live for over a century, a comfortable record not beaten by any other amphibian, even those that are a thousand times bigger.[39]

In this chapter I have teased you with puzzles and heaped you with facts. Here is what it all amounts to. First is the strange fact that for 2.7 billion years, evolution seemed quite content with a world of microbes, devoid of anything multicellular or long-lived. Maybe the extraordinary length of this delay is simply a measure of how intrinsically hard it was for evolution to come up with anything bigger and more complicated, but it is just as likely that the huge evolutionary advantage of short generation time gave microbes a nearly unbeatable edge. We are still battling them. When multicellular condominiums did at last appear, the cells they contained were harnessed and directed to different tasks, providing a means of growing, defending, repairing, and of course most importantly of all, reproducing the organism. This division of labor made longer life possible, but at a price. In animals at least, that price is the risk of cancer caused by rogue cells that behave as though they are parasitic microbes with a life of their own.

At first sight, it seems as though large size and long life go together. For one thing, big animals are better protected against cancer than small ones, but there are lots of exceptions to the correlation between size and longevity. Naked mole-rats and olm salamanders live much longer than their far larger relatives, for example. Even we humans live longer than would be expected for our body size, a puzzle that I will return to. Exactly how long we can live has been the subject of fable and amusing exaggeration. What is certain, however, is the dismal reality that, whether Old Parr or an old parrot, we all degenerate with age.

3
After Many a Summer
AGING

And after many a summer dies the swan.
Me only cruel immortality
Consumes: I wither slowly in thine arms,
Here at the quiet limit of the world,
A white-hair'd shadow roaming like a dream

ALFRED, LORD TENNYSON, "TITHONUS"[1]

Be careful what you wish for, lest fate should answer your call. That is a recurring theme in the myths of ancient Greece. There was once a mortal by the name of Tithonus who was lusted after by Aurora, goddess of the dawn.[2] Aurora had an insatiable appetite for young men and had also seduced Tithonus's brother, Ganymedes. Greek gods seem to have shared all the weaknesses of mortals, almightily magnified. They were promiscuous, jealous, quarrelsome, vindictive, and prone to the high-risk behavior typical of those who do not fear death. They particularly quarreled over their pick of the choicest mortals. Zeus, king of the gods, stole Ganymedes from Aurora, and in compensation the goddess of dawn begged Zeus to give Tithonus, her remaining lover, eternal life. Zeus agreed to the request, though Aurora soon came to regret the gift. As the years passed, Tithonus began to age, his hair turning white, his body shrunken, and his voice whining and shrill.

Too late, Aurora realized that she should have asked Zeus to grant her lover eternal youth, not eternal life. The fate of Tithonus is a reminder of the distinction between aging and longevity. Aging, or more precisely senescence, is the deterioration of biological function through the life span. Senescence limits longevity because it progressively increases the risk of death. Only in myth can senescence and mortality be decoupled from each other. The poet Alfred, Lord Tennyson imagined the decrepit Tithonus's lament to his lover, pleading for release from the cursed gift of immortality so that he might rejoin the "happy men that have the power to die."

If you want long life, then, the thing to wish for is an extension of healthy life span, not just greater longevity. Better be quick about it too, because senescence begins a lot earlier than you might think. Ogden Nash, the American wit who had a verse for every occasion, suggested that

> Senescence begins[3]
> And middle-age ends
> The day your descendants
> Outnumber your friends

Unfortunately, this is overly optimistic. Senescence begins well before middle age, probably soon after puberty, when you become capable of creating descendants and of thinking about life insurance. Though, to be sure, any adolescent who thinks more about death than sex probably needs to consult a shrink more urgently than a financial advisor.

The onset of senescence and its increase during adult life can be tracked through its impact on the mortality rate, typically measured as a percentage (fig. 2). For example, the likelihood of an American male aged 50 dying before his next birthday is

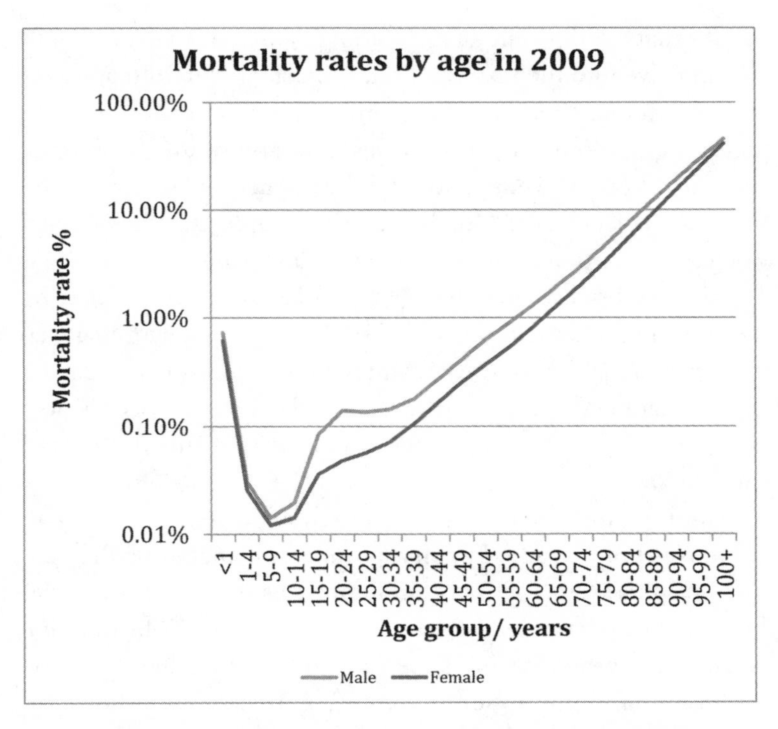

FIGURE 2. Mortality rates by age for the population of the United States in 2009 (measured as likelihood of dying at that age). (Data from World Health Organization.)

about 0.6 percent, according to death rates measured in 2009.[4] As you can see from the graph, the death rate has a peak around birth (neonatal mortality), but the rate then falls and does not rise again until about the age of 15. Thereafter, in all human populations, and indeed in most animal species after reproductive maturity, the risk of death increases with age. How quickly the adult mortality rate increases with age is an indicator of senescence.

The frailties and maladies of senescence are biological and individual, but their aggregate effect on mortality rate turns senescence into a statistical phenomenon as well. Leonard Hayflick, a leading researcher on aging, joked that "it is now proved beyond doubt that aging is one of the leading causes of statistics,"[5] but if you delve into the history of the subject you'll find that this is actually not far from the truth. Senescence was indeed a leading cause of developments in statistics, though not for the reasons one might guess. Today mortality rate statistics are a vital tool used in evidence-based medicine; for example, they helped nail the relationship between smoking and lung cancer. But the first studies of mortality rates and aging pre-date such medical uses by a couple of hundred years. These studies were made for financial rather than medical purposes, and they were motivated precisely by the fact that life is a risky business. The full story reveals how senescence as a statistical phenomenon was first uncovered and measured.

Until the Enlightenment, the church regulated what was and what was not permitted in all aspects of life, including the economic sphere. Among its prohibitions was the sin of usury, defined as charging interest on the lending of money.[6] The doctrine against usury survives in Islamic law today. In medieval Europe usury was anathema to scripture, against Aristotelian philosophy, and held to be contrary to nature. Because interest accrues as time elapses, to earn interest was seen as tantamount to living by selling time itself, which rightly belonged only to God. Though the church and clergy frequently ignored this doctrine when it suited them, usury was not tolerated among the laity. Therefore, those who wished to lend money had to find a profitable method that would not get them into hot water with the church. One acceptable solution was based on the idea that a lender could legitimately be paid for taking a risk with his money. Thus, if the return on a loan was in some way uncertain, earning money by lending could be allowed.

What is more uncertain than life itself? After all, which of us knows the hour of our death? And so life annuities were invented. A life annuity pays you an annual sum of money for the duration of your life in return for a lump sum that you give the annuity provider at the start of the contract. The value of the annuity to you depends on the amount that you invest at the start and how long you live. Obviously, the longer you live, the more money you receive in total, so providers adjust the annuity rate at the start to ensure themselves a profit. An annuity is inherently risky and amounts to a wager between the investor and the provider on how long a person will live. If the person's life is shorter than expected, the provider wins; if longer than expected, the investor wins. For a provider that issues annuities on many lives, making a net profit depends on having accurate statistical information on mortality for the relevant population. Such information is shown in life tables, which are compiled from church or other records of age at the time of death. Mortality statistics are tabulated this way precisely because senescence causes the mortality rate to increase with age. If the information in the life table is accurate and annuity rates are set accordingly, the provider is in the same position as the owner of a casino: a handsome net profit is guaranteed.

Someone should have warned the Scottish poet and lawyer George Outram (1805–1856) that an annuity is a bet and not a sure thing. He shared his own bitter experience with others in a poem written in broad Scots dialect about an annuity that he sold to a recently bereaved widow. Here's a sample of just two of the nineteen verses of Outram's woe-filled lamentation:

> The bargain lookit fair eneugh—
> She just was turned o' saxty-three—
> I coundna guess'd she'd prove sae teugh
> By human ingenuity.
> But years have come, and years have gane
> And there she's yet as stieve's a stane—
> The limmer's growin' young again,
> Since she got her annuity.
> . . .

I read the tables drawn wi' care
For an Insurance Company,
Her chance o' life was stated there,
Wi' perfect perspicuity.
But tables here or tables there,
She's lived ten years beyond her share,
An's like to live a dozen mair,
To ca' for her annuity.[7]

Outram's big mistake was to trust the life tables drawn up for a life insurance company. It wasn't just individual investors who made this error. In the early nineteenth century, the British government lost money heavily when it made two mistakes in issuing annuities to investors. Its first mistake was to use inaccurate data on mortality that overestimated the death rate and hence caused payouts to be made over unprofitably long lifetimes. The misleading data were selected for the government by William Morgan, the actuary of the Equitable Society. His errors had previously made the society money when he used the data to price life insurance policies that paid out on death. Purchasers of such policies were overcharged, since they actually presented a lower risk of death than the society had calculated, and so the society's payouts were lower than expected. The same errors, however, worked in the opposite direction and were financially disastrous when used to calculate the cost of an annuity because, of course, as George Outram discovered, longer lives earned bigger payouts, and consequently the government lost huge amounts of money.

The second costly mistake the British government made was to allow investors to take out annuities on the lives of third parties. This practice enabled investors to stack the odds in their own favor by scouring the general population for individuals who were likely to live longer than the government's data suggested. One of the investors who took advantage of this situation was the poet William Wordsworth, who was in a position to appreciate the longevity of ordinary folk in the Lake District, where he lived:[8]

> Upon the forest-side in Grasmere Vale
> There dwelt a Shepherd, Michael was his name;
> An old man, stout of heart, and strong of limb.
> His bodily frame had been from youth to age
> Of an unusual strength: his mind was keen,
> Intense, and frugal, apt for all affairs,
> And in his shepherd's calling he was prompt
> And watchful more than ordinary men.[9]

Annuities on the lives of the old were particularly profitable for investors because the survival of old people was greatly underestimated by the government, especially in healthier districts, such as the highlands of Scotland, and among certain groups, such as the London Quakers, whom investors discovered were unusually long-lived. Wordsworth profitably invested £4,000 in annuities on the lives of forty old men.

There is an ironic twist to the end of this tale. Until its demise in 2005, the Equitable Life Assurance Society, as the Equitable Society became, prided itself on its long and supposedly distinguished history, using its own longevity as a selling point. But that history contained a warning about the dangers of miscalculating annuity rates that no one heeded and few can have been aware of. The eventual downfall of Equitable Life was brought about by the company guaranteeing some of its policyholders annuity rates that were higher than the society could afford to pay.

For Christian rulers who were in dire need of cash, another solution to the church's proscription of usury was to let Jews do the banking and then to tax or confiscate their profits. In fact, banking was one of the few occupations Jews were allowed to pursue in medieval times, but the profession was no more popular with debtors then than now. This was one of the reasons that Jews were constantly persecuted and then finally expelled from England in 1290 and from Spain in 1492. The presence of a few Jews in England was tolerated by Oliver Cromwell in 1650 expressly for commercial purposes, and over the next century and a half, Jews who had prospered in more tolerant Holland trickled from there into England in pursuit of commerce and finance. In 1779, one Benjamin Gompertz[10] was born into such a family at number 3, Bury Street, in the city of London. For a man destined to revolutionize the study of mortality, fate could not have chosen a more aptly named birthplace.

Benjamin Gompertz was a brilliant mathematician. He was schooled at home, and then self-taught, because as a Jew he was barred from attending English universities. By the age of 19, he was a regular contributor to the *Gentleman's Mathematical Companion*, and when he was older he won the magazine's prize competition 11 years in a row. Gompertz made important contributions to pure mathematics and to astronomy, but it was the mathematical skill that he brought to his day job as an actuary for a life insurance company that earned him a living as well as lasting fame. The job of an actuary is to analyze the statistics associated with various risks and then to use these statistics to calculate the premium that should be charged to insure against them.

Life insurance presents a particular kind of insurance risk because there is no uncertainty about whether a claim will be made, only about when. As the British government discovered to its cost, getting this calculation wrong can be very expensive. The key piece of statistical information is the mortality rate. If people were like glass bottles, the risk of breakage/death each year would be a constant and life tables would not be needed, but humans—and indeed, most animals—senesce, which means that the mortality rate increases with age. The question is, by how much? Benjamin Gompertz found a mathematical answer to this question that is so general that it has been called the Gompertz law.

By studying life tables showing the number of people who die at different ages, Gompertz discovered that from about the age of 20 onward, the mortality rate increases exponentially with age. In other words, the mortality rate doubles at a constant rate. The same phenomenon is seen in other species too, but the rates are different. In humans the doubling of the mortality rate takes about eight years. The Mortality Rate Doubling Time (MRDT) for dogs is about three years, and for lab rats it's about four months.[11] MRDT can be thought of as the rate of senescence. Rats senesce much faster than dogs, and dogs much faster than humans. It is interesting that the rate of senescence appears to be approximately constant for each species. You might think that this is exactly what should be expected from the fact that, as we saw in chapter 2, species have characteristic life spans, but there seems to be a paradox here if we look a little deeper, because life span is not fixed.

Let us take our own species. Two hundred years ago average life expectancy at birth was under 40 years, wherever you lived. Today it is higher than 40 everywhere, even in the poorest nations.[12] In the developed world, people are living longer and longer. In fact, this is not a recent phenomenon. A steady increase has been traced back as far as 1840, since which time life expectancy has increased at the astonishing rate of nearly three months per year, or the equivalent of 15 minutes per hour.[13] The best historical records come from Sweden, where in 1840 women were the world record holders for average life expectancy, although that average was only 45 years, very modest by today's standards. By 2009 female life expectancy in Sweden was 83 years.[14]

Men have shorter lives than women on average, and the gap between the sexes has widened from two years to six, but give or take a few years, remarkable advances in life expectancy for both men and women have taken place throughout the world, especially in richer countries.[15] Life expectancy in the United States has increased especially rapidly since 1970, and whereas the average age at death for men in that year was 67, by 2006 it had reached 75. For women over the same period, life expectancy increased from 75 to 81 years.[16] Very similar increases took place in Britain. In France, the birthplace of longevity record holder Jeanne Calment, women live even longer, and in Japan the average female life expectancy is now more than 86 years. Because these ages are averages, there is, of course, a concomitant increase in the number of people living to still older ages, but at the time of this writing there are no well-documented challengers for Jeanne Calment's crown.

The cause of these remarkable increases in life expectancy is first and foremost a reduction in infant mortality due to improvements in sanitation, obstetrics, public health, immunization, antibiotics, and medical care. All of these and a general increase in the standard of living, which improves the health of the old in particular, have also helped reduce adult mortality. Some of the laggards in the longevity stakes are countries where smoking is especially prevalent.[17] There was an unintentional, mass-scale experiment in human well-being that demonstrated the importance of prosperity to average life expectancy when the Soviet Union collapsed in the 1990s. Economic dislocation and unemployment caused male life expectancy in Russia to fall dramatically by one year per year to a low of only 57 in 1994.[18] This change is a sobering reminder of how quickly the gains of centuries can be lost.

So now we come to the paradox, and it is this. If the rate of senescence, as measured by MRDT, is a constant that in our own species is fixed at a value of about eight years, how is it possible for life expectancy to almost double in only two centuries? Does the increase in human longevity mean that senescence is decreasing? Indeed, as longevity continues to increase, could senescence be progressively conquered, like a disease? Even though he had none of the evidence that we do, Aristotle's penetrating mind led him to wonder something similar. In his book on longevity, he asked whether short life was just a consequence of "unhealthiness" or whether there was some inherent limitation on life span.

The Gompertz law helps us to answer this question and to resolve the longevity paradox.

There are actually two variables in the Gompertz law that determine mortality rates. One is the mortality rate doubling time (MRDT) and the other is the initial mortality rate (IMR), measured at sexual maturity. The IMR can also be thought of as a baseline mortality rate since it actually has an effect throughout life, not just initially. MRDT determines the rate of senescence, but IMR sets its starting point, and of course the higher the starting rate is, the more it will be raised by doubling. To see how these two quantities combine to influence longevity, let's compare two species of birds, the lapwing and the herring gull, that happen to have the same rate of senescence but very different initial mortality rates. In both species MRDT is six years, but the initial mortality rate for lapwings is 20 percent per year, while in herring gulls it is only 0.4 percent, or fifty times smaller. Although both birds senesce at the same rate, the oldest lapwing ever recorded in the wild was only 16 years old, while the oldest herring gull was 49.[19] Since the two species senesce at the same rate, the difference in their longevities must be due to their very different IMRs. Notice that although the difference in IMR between the species is fiftyfold, the difference in their maximum longevities is only threefold (16 years vs. 49 years). This is because MRDT (which is the same in the two species) has a stronger effect than IMR. Senescence rules in the Gompertz law.

The comparison of the lapwing and the herring gull illustrates the point that we should not jump to the conclusion that an increase in human longevity is caused by a decrease in senescence. MRDT is fairly constant across human populations in both the developed and the developing world, while initial mortality rates differ greatly and have changed a lot. The truth is, then, paradoxical though it sounds, that while human longevity is increasing, we continue to senesce at the same rate. The explanation is that increasing longevity is caused by a decreasing initial mortality rate, not by a decrease in senescence. If we hypothetically removed senescence so that the mortality rate did not increase at all after age 20, then humans could easily live as long as Methuselah.

In effect, increasing longevity is the result of senescence being delayed, not reduced. So we now have an answer to Aristotle's question of whether length of life is subject to some inherent limit. The obduracy of senescence tells us that unless the aging process itself can be slowed, it ultimately places a statistical limit on length of life. Rising life expectancy also suggests, however, that we have not reached that limit yet. The scientific literature on human longevity is littered with predictions that turned out to be far too pessimistic. In 1928, for example, Louis Dublin used US statistics to calculate a best-case scenario for life expectancy "in the light of present knowledge and without intervention of radical innovations or fantastic evolutionary change in our physiological make-up, such as we have no reason to assume." His prediction was that maximum life expectancy would be 64.75 years. Though Dublin was unaware of the fact at the time, women in New Zealand had already achieved a higher average life expectancy than this.[20] Dublin and others later revised their predictions upward more than once, and each time the inexorable rise of longevity burst through its supposed limits. By one estimate, if present trends continue, the majority of children born since 2000 in the richer countries can expect to live to 100.[21]

As you will have noticed, aging is a frequent subject for poetry. The poet Henry Reed, surely unwittingly, stumbled across a rather neat summary and resolution of the longevity paradox in his satirical poem "Chard Whitlow," written in 1941 as a parody of T. S. Eliot:[22]

> As we get older we do not get any younger.
> Seasons return, and today I am fifty-five,

And this time last year I was fifty-four,
And this time next year I shall be sixty-two.

Apparently T. S. Eliot enjoyed the parody. In "The Love Song of J. Alfred Prufrock," he makes the equally (and intentionally) fatuous-sounding observation: "I grow old . . . I grow old . . . / I shall wear the bottoms of my trousers rolled." The human frame shrinks with age, which is why an old man wearing an old pair of trousers would be well advised to roll them up. And as for Henry Reed's leap from 55 to 62 in a single birthday, senescence effectively means that biological time really does speed up with the passing years. Up to a point, that is, for a surprise lurks in the depths of very old age.

As longevity has risen, more and more people are living into their second century, and these pioneers at the frontier of survival have afforded us a glimpse into the hitherto uncharted territory of extreme old age. The news from the other side of the century line is better than many dared hope. Unlike Tithonus, whom age cursed with increasing debility, a proportion of centenarians have surprisingly good health. For example, one-third of a Danish group of centenarians were healthy enough to live independently.[23] Even more remarkably, 40 percent of a group of American supercentenarians between the ages of 110 and 119 were healthy enough to live independently, or required only minimal assistance to do so.[24] And if you happen to be a lab mouse belonging to a strain that has been bred for unusually long life, you too will be healthier in extreme old age than your shorter-lived ancestors.[25] The explanation is surely a simple one: good health is the key to reaching old age, whether mouse or human. But there is another discovery from the land of the very long-lived that really is a surprise: when you get really old, senescence ceases.

Estimating mortality rates among the oldest old is difficult because until very recently there were so few individuals who qualified. Finally, in 2010, a study that accumulated mortality data on over six hundred genuine supercentenarians showed what had been suspected for some time: that the mortality rate in this group comes to a standstill.[26] To be sure, the mortality rate is very high at this age, and 50 percent die every year, but the mortality rate does not increase any further each year. I suppose this result might be greeted among supercentenarians like a good news/bad news joke: "The good news is that you have stopped aging, the bad news is that you are going to die soon anyway." From a scientific point of view, it is certainly of great interest, but what can it mean? With so few supercentenarians available for study, the answer must come from another species and from an unexpected direction.

In the south of Mexico, a stone's throw from the border with Guatemala, is a factory that produces something rather unusual: 500 million pupae of the medfly every week. The medfly is a damaging pest of citrus fruits, but the huge numbers of flies bred in the Mexican factory are part of the solution, not the problem. Female flies mate only once, so the strategy used against them is to swamp an infested area with huge numbers of males that have been sterilized in the rearing facility. Where sterile males outnumber wild ones, females mate with the sterile males but produce no offspring. The task of the medfly-rearing facility in southern Mexico was to prevent the spread of the medfly northward through Mexico and into the United States, and it was spectacularly successful.[27] The use of medflies from the facility in aging research was just a spin-off, made possible by the huge number of flies available.

Aging researchers followed the fate of 1.2 million medflies reared in the Mexican factory, a mere 1 percent of the flies produced there on any day of the week, but a huge sample for such a study. Medflies, like most insects, are short-lived as adults. After 7 days the daily mortality rate was a modest 1.2 percent; two weeks later it was nearly 10 percent. After 40 days there were 45,000 old flies left, and the mortality rate was 12 percent a day, but then it began to fall. After 90 days there were just a hundred or so flies remaining alive, but their mortality rate was down to 5 percent a day, and it was another 82 days before the last fly died.[28] This result was even more dramatic than that later found in human supercentenarians, for among medflies the mortality rate did not just stall, it went into reverse, as the mortality rate of the oldest old flies actually declined with advancing age. At first sight, one might think that these studies show that even senescence gives in to old age eventually, but there is another interpretation.

In flies, as in humans, it is the frailest, least healthy individuals who die first. Many factors affect health and might differentiate those fated to live long from those destined to die early. Among those factors is sex. Females live longer in humans and in many other species, but not all. Males live longer than females in rats, for example.[29] Whatever the cause of differences among individuals' life chances might be, the mere existence of such differences can generate the appearance of a declining mortality rate for the population as a whole.[30] This happens because as the shorter-lived group dies out, the more robust individuals with the lower mortality rate are the sole survivors. In such a situation, has the mortality rate really declined, or has the lower mortality rate that was always present in a fraction of the population just been uncovered by the deaths of shorter-lived individuals?

Imagine an analogous situation where we have a bathtub in which are floating equal numbers of blue balls and yellow balls. Both types of balls absorb water and eventually sink out of sight, but the blue ones do so at a slightly faster rate than the yellow ones. Let's step back and watch what happens. From a suitable distance, the mixture of blue and yellow balls merges into green, like the pixels on a TV screen or in a newsprint photograph. Then, as the balls begin to "die," the surface of the bathtub appears to change color. The blue balls sink faster than the yellow, and as the blue ones disappear beneath the surface, its color slowly changes from green to yellow. Near the end the surface of the tub is all yellow. What has happened? Have the contents of the tub really changed color from green to yellow, or has differential mortality among balls merely revealed something that was there all along? Half the balls were yellow, but we could not see this because, at the population level, we perceived the mixture as green.

Now let's forget the colors of the balls and just look at the rate at which the balls in the tub sank, or "died" if you prefer, during the experiment. Since there was a mixture of slow- and fast-sinking balls at the start and only slow-sinking balls were left at the end, the average rate of sinking slowed toward the end of the experiment. How should we interpret this observation? One way is to believe that all the balls were the same (we are ignoring their color, remember), which would mean that they must have changed their susceptibility to sinking during the experiment. Another way is to believe that the balls were not all the same and that they differed in their mortality rates all along.

Except in rare situations, biological populations are not composed of identical individuals. They are composed of individuals that differ in all kinds of ways, many of which affect health and mortality. In such a situation the composition of any population will change, especially toward the end of the life span, and this can give the impression that the mortality rate has stalled, or even begun to decline. A likelier explanation is that the population contains hidden differences in mortality risk. This is a very exciting possibility, because if we can only find out what these differences are, we may be able to get to the reason why some individuals live longer than others.

Let's now recap what we have discovered about senescence. Senescence is the gradual loss of biological function that occurs with age. There is a joke that old university professors never retire, they just gradually lose their faculties. In his comedy *As You Like It*, William Shakespeare had the melancholy traveler

Jaques describe the last of the seven ages of man as a "second childishness and mere oblivion. Sans teeth, sans eyes, sans taste, sans everything."[31] Trends in senescence can be tracked through its effect on the mortality rate. Benjamin Gompertz discovered that once sexual maturity has been reached, the mortality rate increases exponentially, with a characteristic doubling time of about eight years in humans. Although in the richest countries life expectancy has approximately doubled over the last 200 years, the mortality rate doubling time has not declined. The explanation for this paradox is that senescence has not been reduced; it has just been postponed to later life.[32] We have no idea what further gains in life expectancy may be made in the future, but we can say that such gains are unlikely to be made at the expense of senescence. Senescence does come to a stop in extreme old age, but by then the annual mortality rate is so high that this buys very little extra time.

The grinding to a halt of senescence in the very old is probably caused by death winnowing out the frailest, leaving behind those who have enjoyed more robust health than average throughout their lives. Could such good health in old age be inherited? Advances in genetics have been so rapid in the last 20 years that we can now see directly the genetic differences among individuals. What do these differences tell us about aging?

4
The Eternal Thing
HEREDITY

The years-heired feature that can
In curve and voice and eye
Despise the human span
Of durance—that is I;
The eternal thing in man,
That heeds no call to die

THOMAS HARDY, "HEREDITY"

The nineteenth-century American physician and writer Oliver Wendell Holmes was popular in his day for his conversational essays, written as if the reader were sharing breakfast with the author. So personal was the connection he made with his *Breakfast-Table* readers that they would frequently write to him for advice, to which requests he responded when he turned 80 in a book called *Over the Teacups*.[1] Life, he seemed to imply, is a mere interlude between breakfast and tea. Holmes was asked how he had achieved his feat of longevity and how his correspondents could do likewise. Even today, 80 exceeds the average life expectancy of an American male, so back in 1889, reaching "three score years and twenty," as Holmes liked to call it, was a venerable achievement.

"One of my prescriptions for longevity may startle you somewhat," he wrote. "It is this: become the subject of a mortal disease. Let half a dozen doctors thump you, and knead you and test you in every possible way, and render their verdict that you have an internal complaint; they don't know what it is, but it will certainly kill you by and by." Then act the invalid, Holmes advises, nurse your mortal complaint as if it were a baby, and you will probably make it to 80. And if you do, you will find that most of your friends are dead and that while you were worrying about your health, life has passed you by. So take advance precautions, Holmes advised, and "some years before birth . . . advertise for a couple of parents both belonging to long-lived families."

Longevity does seem to run in families, but how much and why? Oliver Wendell Holmes's son, Oliver Wendell Holmes Jr., appears to have followed his father's advice, if not his wishes, to the letter. Starting with a judicious choice of long-lived parent, Junior lived life to the full, interrupting his college studies to volunteer in the Union army when the American Civil War broke out. Despite being wounded three times, he survived the war and ultimately became an associate justice of the Supreme Court, where he served until the age of 90. I feel a small personal connection with this story because my own father was a lawyer who chose to give me the middle name Wendell in honor of the judge.

I also chose my father well, since at the time of this writing he is 98 and a model of good health in old age. He survived diphtheria as a child, an often fatal bacterial disease that was common until widespread vaccination began in the mid-1920s. Later, during the Second World War, he survived torpedoes and shipwreck. Clearly, surviving to an advanced age involves a certain amount of luck, helped, as my father insists, by an ability to swim. He still swims three times a week, though these days without his Navy-issue tin hat.

For the lucky survivors, what part do genes play in reaching a healthy old age? Many studies have addressed this question by comparing the longevity, health, and, indeed, the DNA of the long-lived. A ballpark estimate that seems to apply to mice, nematode worms (a favorite of aging researchers), and humans is that genes account for between 25 and 35 percent of the variation in how long individuals live.[2]

Nearly all important characteristics that vary among individuals are affected by both genes and environment, and teasing the influences of the two apart is tricky, if not to say sometimes controversial. We know that genes set the potential for long life, but not its absolute limit, from the fact that human life expectancy has doubled in the last 200 years solely due to improvements in public health, medicine, and prosperity. Among animals too, environmental influences can greatly affect life span. A queen honeybee lives and reproduces for several years while tended by workers who are her genetically identical sisters, but who live only a few months.[3] The different fates of genetically identical queen and worker are set during early development. Workers caring for the brood feed selected larvae on a pure diet of a protein-rich secretion called royal jelly, and these larvae become queens. Larvae fed only a small amount of royal jelly at the end of development become workers. Needless to say, the Internet merchants selling royal jelly for its supposed anti-aging properties omit to mention that you need to be a bee larva of a certain age to benefit. Neither do they issue a health warning that taken in too small a dose it might turn the recipient into a worker.

In humans, estimates of the genetic contribution to a trait such as life span are made by comparing variation between identical twins with variation among non-identical siblings. Identical twins arise from a single fertilized egg, called a zygote, that splits in very early development. They are more accurately called monozygotic (MZ) twins, because even though they are genetically identical, MZ twins are not truly identical in every other way.

Twins are usually raised together in the same family, so they experience the same environment as well as sharing the same genes. This makes it tricky to work out whether certain similarities among MZ twins are due to environment, genes, or a combination of the two. Luckily, there is a way around the problem by comparing MZ twins with dizygotic (DZ) twins. DZ twins are not genetically identical to each other, though like MZ twins they are born and usually raised together. Twin studies illustrate that the influence of genes on aging and longevity is not nearly as neat and tidy as a figure like "up to 35 percent" makes it sound. The chance of your being diagnosed with Alzheimer's, for example, is two to three times greater if your MZ twin has been diagnosed with the disease than if you have a DZ twin with Alzheimer's. Although this observation demonstrates that the risk of getting Alzheimer's is very significantly affected by your genes, the age of onset of the disease can vary by many years between MZ twins, and some escape the disease altogether, demonstrating that nongenetic influences are also important. An exception occurs in the rare form of Alzheimer's (5 percent of cases) called early onset familial Alzheimer's disease, in which the presence of specific genetic defects invariably leads to the occurrence of the condition before the age of 60.[4]

If we were lab rats in lab coats (*Rattus norvegicus*), we'd want to study humans as models of aging, so wonderful is the species *Homo sapiens* as an experimental subject, especially those individuals living in Nordic countries (*H. sapiens norvegicus*), where health is good and records are well kept. A nearly complete sample of all the twins born in Denmark, Finland, and Sweden between 1870 and 1910 was used to study the effect of family history on age at death in 20,502 people.[5] For individual twins who died before the age of 60, there was no correlation with the life span of the survivor, whether monozygotic or dizygotic. In other words, shared genes did not influence deaths before 60 years of age, and environmental factors had an overriding effect on longevity up until this age. After 60, however, twins' ages at death were correlated, revealing an effect of shared genes that grew stronger with advancing age. The same general pattern was true for both men and women, though women lived longer on average than men.

Recall from chapter 3 that senescence slows at very advanced ages in populations of humans and other species, and that a likely explanation for this pattern is that populations are in fact composed of multiple groups that senesce at different rates. The study of Nordic twins supports this idea, since it also suggests that there is something genetically different about people who reach older ages. Do the genes that are favorable to longevity operate at younger ages too? The absence of genetic effects on longevity below the age of 60 in the Nordic study might suggest not, but such effects might just as easily be hidden by the fact that mortality from all causes is lower below 60. One way to find out is to compare the health of middle-aged offspring of parents who have reached 90 and above with the health of middle-aged offspring whose parents died before this age. Studies of this kind on the middle-aged offspring of centenarians and supercentenarians have indeed shown they that are healthier than the norm, but this might just be because they were taught as children to follow the healthy lifestyle that enabled their parents to live so long. It might have nothing to do with genetics. A study at Leiden in Holland found a way to test this out.[6]

The Leiden Longevity Study tracked down families in which there were two or more siblings who had reached the age of 90 and compared their mortality rate, state of health, and the health of their middle-aged offspring with that of randomly selected nonagenarians and their offspring. One 90-year-old in a family might have reached that age through chance alone, but the likelihood that two or more siblings could have reached such an advanced age by chance, without the help of genes for longevity, ought to be much less. And so it proved, for the nonagenarian siblings had a mortality rate that was 40 percent lower than the randomly selected nonagenarians, supporting the hypothesis that the siblings were genetically predisposed to long life. Not only that, but their parents and their offspring also had lower mortality rates than the general population.[7]

Next, the study compared the health of offspring of the genetically favored parents with the health of those offspring's partners to find out whether inherited good health was evident in offspring during middle age. The logic behind this comparison was that two partners would share the same lifestyle and environment, but would not, except by a very remote chance, both come from long-lived families. So, if having nonagenarians in the family is a sign of inherited good health, this should show up in such a comparison. The differences between offspring and their partners were relatively small, but did show the expected effect of better health in the middle-aged offspring of nonagenarians, who had lower risks of heart attack, hypertension, and diabetes mellitus than their partners.[8] Other studies have come to a similar conclusion: exceptional longevity is clustered in families whose members enjoy better than average health throughout life.

These studies tell us that there must be genes that predispose to long life in humans, and that making it to 90 or more is not simply due to good fortune or a favorable environment, as important as these are. So, what are the longevity genes? This question is where a lot of the action is in gerontology at the moment, because genes are like switches that, at least in theory, can be flipped from a less desirable state to a more desirable one to bring the benefits of health and long life to those not fortunate enough to have inherited them. But before you can flip the right switch, you have to find it in a maze of genetic circuitry.

The first longevity gene was discovered in a minuscule nematode worm with a name that is thirty times longer than its body: *Caenorhabditis elegans*. This tiny creature, only 1/32 inch long, is the Solomon Grundy of gerontology, living as brief a life as the nursery rhyme character who was "Born on Monday, Christened on Tuesday, Married on Wednesday, Took ill on Thursday,

Worse on Friday, Died on Saturday, Buried on Sunday." And the ecology of *C. elegans* in the wild is scarcely better known than the short career of Solomon Grundy. The worms are found in soil, where they feed on bacteria, although which kind of bacteria they prefer is not known. They are mostly hermaphrodites, the majority of worms having both male and female sex organs. If food is scarce or the environment is overpopulated with *C. elegans*, young worms enter a stage of arrested development called a "dauer" (German for "enduring"). Dauers, like seeds in plants, are good at dispersal as well as endurance. Dauers have been found attached to the small invertebrate animals that live in soil, such as snails, slugs, mites, and millipedes.[9] And that's it. When Solomon Grundy's obituary is published on Monday in the *Worm Breeder's Gazette*,[10] there is little more to say—unless you want to know about his genetics, that is. Of that there are volumes and volumes of biography.

The normal life span of *C. elegans* in soil is only a few days. Unhampered by the need to find and court a mate or to care for its family, a hermaphrodite worm born on Monday is the single parent of a couple of hundred offspring by Solomon Grundy's wedding day (Wednesday).[11] The lives of worms on a culture plate in a lab, however, are much longer, and in this protected environment they can live for three weeks and be bred in experiments. In the 1980s breeding experiments with mutant worms managed to increase life span significantly, demonstrating the influence of inheritance and therefore the existence of genes for length of life.

The first longevity gene was identified by Tom Johnson and David Friedman at the University of California, Riverside.[12] The gene, which they called *age-1*, increased the average life span of worms carrying it by a massive 65 percent, mainly due to a decrease in the rate of senescence (an increase in MRDT).[13] The *age-1* gene turned up in three different long-lived mutants, leading Johnson to suggest that this was the only gene for longevity in the worm, but the story soon became more complicated.

As the captive worm ages, it is normally lost to view in the anonymous crowd of its rapidly multiplying descendants, but Cynthia Kenyon, another leading scientist in the genetics of aging in *C. elegans*, describes what happened one day back in the early 1980s when she first saw an aged worm:

> I left culture dishes with my almost-infertile mutants in the incubator for several weeks, and then looked at them. With so few progeny, the original animals were still easy to find, and to my surprise, they looked old. This concept, that worms get old, really struck me. I sat there, feeling a little sorry for them, and then wondered whether there were genes that controlled ageing and how one might find them.[14]

And as so often occurs in science, a chance observation and the curiosity it provoked led to discovery. When Kenyon and her team started to screen mutant *C. elegans* for exceptional longevity, they quickly came across a strain with a mutation in a gene called *daf-2* that lived more than twice as long as normal worms. This same gene had long been known to influence dauer formation, an event that takes place during juvenile development, but now it seemed that it also acted in adult worms to increase their powers of endurance too. Then the team discovered that another gene involved in dauer formation, called *daf-16*, was also involved in life span extension. In its normal, unmutated state, *daf-2* keeps *daf-16* switched off, and worms have a normal life span. Mutation in the *daf-2* gene inactivates its effect on *daf-16*, and this gene is switched on and lengthens life.[15] It later transpired that *age-1* also affected life span through an effect on *daf-16*.

Genes operate in concert, not alone. The light switch in the corner of the room is part of a circuit and functions only when connected to a source of power and to a lightbulb. Likewise, a gene that switches life span from normal to double-normal has got to be connected to some mechanism that is responsible for the change. Finding the first switch was very important because it proved that a mechanism for extending life span must exist. Before that, it was as if biologists had been living in a permanently darkened room, unable to imagine light or to conceive of a switch that could manipulate life span. The prevailing idea was that organisms just wear out. The discovery that flipping a genetic switch lengthened life was therefore brilliantly illuminating.

Once the switch had been found, the big question was, what is the mechanism that is being operated by the switch? While it is useful to think of genes as switches, they are really more than that. Like an electrical switch in a circuit, a gene has a position in a pathway, and like an electrical switch, a gene makes or breaks links. The links in a biochemical pathway, however, are forged by molecules, not by wires conducting electrical current. Those molecules are produced directly or indirectly by the genes, so reading the DNA sequence of a gene can tell you which molecule it produces. Hence, whereas looking at the construction of an electrical switch will tell you almost nothing about the circuit it controls, looking at the DNA sequence of a gene can tell you a great deal. So in 1997, when the *daf-2* gene was decoded, there was another surprise in store.

The DNA sequence of the *daf-2* gene identified it as a switch that is triggered by the worm version of the hormone insulin.[16] Subsequent research rapidly discovered that the equivalent insulin signaling pathway (insulin → *daf-2* → *daf-16*) is also present in yeast, fruit flies, and mice, and that inactivating the *daf-2* gene also lengthens life in these species.[17] Evolution apparently produced a pathway to longer life over a billion years ago and has conserved this pathway among eukaryotes to this day. The conservation of this pathway must mean that it has an important function, but what is it? Its primary function cannot be simply to lengthen life, because if that was always advantageous, then naturally occurring *daf-2* mutants with longer lives would be the norm.

Anyone who lives with diabetes will be familiar with insulin and its role in regulating levels of glucose (sugar) in the blood. Glucose is the circulating fuel that powers cells, but like having the fuel in your car spill out of the engine, it is dangerous to have glucose present in the blood at too high a concentration. Type 1 diabetes is caused by insufficient production of insulin in the pancreas, and it is routinely treated by regular injections of insulin, which reduce blood glucose levels. Type 2 diabetes is caused by the cells that use glucose—such as fat, muscle, and liver cells—becoming insensitive to insulin, which normally triggers their uptake of glucose. Type 2 diabetes is often associated with obesity and can usually be corrected by increased exercise and a change in diet, but there are a few patients with type 2 diabetes in whom the condition is caused by a genetic mutation in the gene that insulin normally activates. This insulin receptor gene is the human version of *daf-2*. Despite the billion years of evolution that separate nematode worms and mammals from each other and from their common ancestor, there is a functional part of the worm *daf-2* that is 70 percent similar in its genetic code to the equivalent part of the human insulin receptor gene, although the genes have diverged in their functions.[18]

The function of the insulin signaling pathway in *C. elegans* is a big clue to its role in animals more generally. In its natural, unmutated state, called the "wild type," the insulin signaling pathway determines whether the development of a young worm will follow a direct course from small larva through several intermediate stages to adult, or whether the larva will instead enter a state of arrested development as a dauer. Dauers cannot feed, but they can persist for long periods before they resume development to adulthood. Dauer development is triggered by food shortage and crowding, which worms detect through sensory organs in the head and tail. The role of the sensory organs is demonstrated by the fact that mutant worms with disabled senses have an extended life span as well as a much greater propensity to form dauers.[19]

Thus, the function of the insulin signaling pathway in *C. elegans* is to use sensory information on the condition of the environment to channel development down the most appropriate route. In good conditions, when food is plentiful, worms reproduce, and die as a consequence shortly afterward, but in poor conditions they become dauers and sit it out till conditions improve. Mutations in the genes that induce dauer development also, perhaps coincidentally, lengthen adult life span. Exactly how this happens is not known, but mutations affecting insulin signaling have similar life-lengthening effects in yeast, worms, flies, and mice, in all of which mutants are better protected against a whole range of the hazards of aging, including cancer.[20]

It is paradoxical that a mutation in the insulin signaling pathway that lengthens life in a whole variety of species is the cause of diabetes, and thus threatens life, in humans. Why this should be is not properly understood, but a likely explanation is that there are optimal levels of insulin signaling, and that the right amount varies among species and even among different tissues.[21] Each organism has two copies of every gene, one inherited from each parent (including self-fertilizing hermaphrodites like *C. elegans*, which have two parents in one). Even in *C. elegans*, in which a mutation in one copy of the *daf*-2 gene extends the life span, the effect is lethal if both copies are mutated. Furthermore, the insulin signaling pathway is more complex in mammals, as it involves additional genes triggered by insulin or insulin-like hormones, not just the one found in worms and flies. One of these genes is triggered by a hormone called insulin-like growth factor 1 (IGF-1). Mutations that inactivate one of the two copies of the gene that is switched by IGF-1 are associated with longer life in humans and mice.[22] This observation suggests that genes involved in the hormonal regulation of nutrients (glucose) and growth can lengthen life span in all species, including our own.

Another gene that seems to have nearly universal importance in the regulation of life span in response to available nutrients and energy is the gene that produces a protein by the name of TOR, which stands for "target of rapamycin." The story of how it got this rather odd name sheds light on its evolutionary history. Rapamycin is an antifungal compound produced by bacteria that were discovered living in a sample of soil taken from the remotest inhabited island on the planet—Rapa Nui, also known as Easter Island. Rapamycin is produced as part of the chemical arsenal that microbes use against one another. Firing back in the other direction, the antibiotic penicillin is an example of an antibacterial compound produced by a fungus. The compounds that bacteria and fungi use in their war of the micro-worlds are often finely targeted at the enemy's vital functions. So when rapamycin was added to experimental cultures of yeast, which is a fungus, it was not surprising that most cells were poisoned. What was surprising was that a few cells were unaffected. The yeast cells that were resistant to poisoning by rapamycin carried a mutation in the gene that produces the target of rapamycin. This gene was subsequently shown to be performing a vital function, not only in single-celled organisms like bacteria and yeast, but also in all multicellular organisms, including plants, worms, flies, and mammals.

With the benefit of hindsight we can see that, despite the bizarre history of its discovery, TOR was likely to be vitally important because it was a target of microbial warfare, in which weapons aimed at vital targets are naturally selected. We can also see that it was likely to be found in multicellular organisms because pathways that perform vital functions are conserved by evolution. But what vital, universal function does the *TOR* gene have? The answer is that it controls the growth of cell size in response to the availability of raw materials such as amino acids and the presence of signaling molecules like IGF-1.[23]

So *TOR* is another important gene involved in controlling the balance between cellular growth and maintenance, and like the insulin signaling pathway, it can be manipulated to alter life span in all the usual laboratory suspects: yeast, worms, flies, and mice. Experimental feeding of rapamycin to 600-day-old mice increased their life span by about 10 percent.[24] In humans this would be the equivalent of adding five years to the life expectancy of a 50-year-old. Through its effects on TOR, rapamycin seems to be able to reverse some of the detrimental effects of aging. The clearest evidence of this so far comes from experiments with cells taken from patients with a very rare inherited condition known as Hutchinson-Gifford progeria syndrome, or just progeria for short.

Progeria is a disease occurring in about 1 in 4 million births, caused by a mutation in just one particular gene. Babies appear normal at birth, but suffer retarded growth and develop conditions such as hair loss, wrinkled skin, and hardening of the arteries (atherosclerosis) that are normally associated only with people in their sixties or older. Children with progeria commonly survive to only 13 years of age, most often dying of heart attack or stroke. Despite the obvious similarities to aging, progeria is not simply an acceleration of the normal aging process. For one thing, progeria is caused by a mutation in a single gene, while hundreds of different genes are associated with normal aging.[25] It has been discovered, however, that rapamycin can reverse defects in cells taken from progeria patients.[26] This discovery offers hope, not only of a treatment for this dreadful disease, but perhaps also of a route to ameliorating some of the effects of normal aging on cells.[27] Unfortunately, rapamycin itself is not suitable as an anti-aging drug because it suppresses the immune system, and is in fact used for this purpose in organ transplantation.

Many genes are now known to influence life span in laboratory experiments, though their significance, if any, in humans is often obscure. One gene that does seem to have a significant effect in humans is *APOE*. This gene is central to how the body processes low-density cholesterol and fats, and it has at least seven variants, or alleles, that have differing effects on a wide range of age-associated diseases as well as on longevity. In populations of European descent, the three common *APOE* alleles are those called epsilon-2 (ε2), ε3, and ε4. People carrying two copies of ε4 are at greater risk of heart attack (cardiovascular disease) and death as they get older than are those with other alleles.[28] However, people with the allele combination ε3/ε4 experience some protection from cancer, which balances out the extra risk of death from heart attack.[29]

Choosing long-lived parents who can endow you with longevity genes, as Oliver Wendell Holmes Sr. advised, is fortunately not the only way to longer life. Benjamin Franklin (1706–1790) recommended a much more practical solution: eat less. This advice was based on the personal account of Alvise (know as Luigi) Cornaro, a sixteenth-century Italian entrepreneur who made his fortune early in life and plowed his wealth into self-indulgence. At the age of 35, no doubt corpulent and certainly exhibiting symptoms that would today indicate type 2 diabetes, he was told by his doctor that a radical change of lifestyle was needed or he would not live a year longer. At once, Luigi changed his eating and drinking habits, taking as his guideline never to eat his fill, but always to leave the table before his appetite was satisfied. In a tract called *Discorsi de la vita sobria* [Discourses on the Sober Life][30] that he later wrote advocating this way of life, Cornaro reported that he immediately began to feel better, and that within a year he was fully recovered and in the peak of health. His mainstay was a low-calorie soup, known in Italian cuisine today as *panado*, plus two glasses of wine a day. This diet probably provided him with between 1,500 and 1,700 calories a day,[31] significantly short of the 2,000 calories that are nowadays recommended as an adequate diet for a man. Today we would say that Luigi Cornaro practiced what biologists describe as dietary restriction (DR), consuming a diet that was just short of malnutrition.

Cornaro lived to the ripe old age of 83, more than twice the average life expectancy that prevailed anywhere during his day.

His book was translated into several languages and eventually found its way across the Atlantic, where it was published in 1793 with an essay on good health by Benjamin Franklin and a blurb by President George Washington, no less.[32] He was not the discoverer of the health benefits of eating little, for after all, he had merely followed his doctor's orders, albeit with greater dedication than many of us can muster, but he certainly provided a role model for *La vita sobria*. This diet evidently cured Cornaro's diabetes and would certainly have dramatically lowered his blood sugar and insulin levels.

There are people who follow Luigi's example today, limiting their diet with equally ferocious determination. Some of these people are being studied, but the jury is still out on whether extreme caloric dietary restriction, as distinct from dieting just to avoid overweight, lengthens human life. Extreme caloric restriction has side effects too. People practicing it feel perpetually cold, are understandably short of energy, and have low libido,[33] symptoms that have an uncanny resemblance to the dauer state of *C. elegans*. Personally, I agree with Woody Allen, who has said, "You can live to 100 if you give up all the things that make you want to live to be 100."[34] The remarkable thing is that caloric DR does significantly lengthen life in gerontologists' experiments with most of their favorite model species, from yeast through *C. elegans*, fruit flies to lab rats, although two different studies with monkeys have produced results that contradict each other.[35] The picture is not always clear as to what the genetic pathways are that link DR to longer life, and these pathways seem to vary between species. Even so, the usual suspects are often implicated as the ultimate target: insulin and insulin-like signaling.[36]

I now want to invoke my inner Wendellness and make a Holmes-like suggestion of my own. When O. W. H. Senior recommended, over his teacups at his fashionable address on Beacon Street in Boston, that anyone wishing to live to 80 should "advertise for a couple of parents both belonging to long-lived families," he was indulging in too modest a fantasy. Let those parents both belong to families of conifers, and you can live for 4,000 years or more. Plants such as the famous bristlecone pines of California are the ultimate examples of long life. In fact, one might wonder, do they even senesce at all?

5
Green Age
PLANTS

> The force that through the green fuse drives the flower
> Drives my green age; that blasts the roots of trees
> Is my destroyer
> And I am dumb to tell the crooked rose
> My youth is bent by the same wintry fever
>
> DYLAN THOMAS (1934)

In his poetry Dylan Thomas attempted to show the relationships between his subjects by combining, even colliding, their images. In a letter to another poet, he wrote, "Show, in your words and images how *your* flesh covers the tree & the tree's flesh covers you."[1] Science also seeks the underlying unity of nature, but it is alert to the differences too. Remote mountain valleys shelter a paradise of human longevity only in the imagination of novelists and the wishful thinking of health nuts, but the Shangri-La of human fantasy is a reality for plants. Any botanist in search of long-lived trees can trek high into the White Mountains of California to pay homage to the bristlecone pines named Methuselah and Patriarch in ancient groves. The oldest living bristlecone was an extraordinary 4,789 years old when its trunk was cored to count the annual growth rings in the middle of the last century.

An even older one found in Nevada was chopped down to retrieve the corer used to age it when the tool broke in the hands of an overeager student.[2]

Most of the oldest trees in North America are westerners, but there is a remarkable exception. An eastern white cedar growing unchecked in normal forest is a short-lived tree that matures in 80 summers, but rappel down the face of the Niagara escarpment in Ontario, Canada, to where gnarled specimens of the same species grow, and you will find eastern white cedars with 1,800 annual rings.[3] When the tree lives in a lofty rock crevice, subjected to the rigors of drought, scant soil that offers only short rations, and the hazards of icefalls and rockfalls that sever roots and branches like an unforgiving trainer of bonsai, cedar time slows almost to a halt. It is as if the wind that howls around the almost-naked cliff face, where cedars cling as precariously to life as to rock, whispers Luigi Cornaro's ghostly motto: "Eat little—live long."

There certainly seems to be a general correlation between slow growth and long life. The bivalves that reach nearly half a millennium in the cold waters of the North Atlantic grow slowly and surely. The ancient bristlecone pines that look their age—with scant foliage, weathered trunks, and twisted branches curtailed by a cruel climate—grow slowly too. What these organisms, bivalve and bristlecone alike, have in common is that their growth is indeterminate: however slowly they grow, the capacity for growth never ceases.

While indeterminate growth is rare among animals—confined to sea creatures such as certain fishes, lobsters, corals, and mollusks—it is almost universal in plants. Growth in plants and corals is indeterminate because of the particular way they are constructed. Each is made up of a series of connected modules; these modules are shoots in plants and zooids in corals. Every module is capable of growing additional modules that can add to the size of the plant or coral and that can replace dead ones.

As a consequence, the oldest corals, which live in deep water, can live for thousands of years.[4]

It is clear that Aristotle must have understood the importance of indeterminate growth when he wrote that the long life of plants is due to their capacity to renew themselves. However, the ancient Greek philosopher would not have known how plants manage this. Strange to tell, the bulk of a tree is dead. In the trunk, only the outermost layers of cells beneath the bark are alive. Just beneath the bark is a layer called the phloem, which conducts sugars from the leaves where they are made downward to the roots. Beneath the phloem layer is a layer of dividing cells called the cambium. These dividing cells are responsible for producing the phloem on the outer surface of the cambium layer; on the inner surface they divide to make a layer called the xylem. The cells of the xylem fulfill their function mainly in death. When dead, they form hollow vessels that join end to end into continuous pipes that move water from the roots to the leaves and every other living part of the tree.

The rate of cell division in the cambium is at the mercy of the environment, and so it varies with the seasons and from year to year. In temperate regions, cambial cell division is at its most vigorous in spring, when it produces large-diameter vessels. As the months pass, the temperature falls, and water becomes less available, the new xylem vessels produced are narrower and narrower, until growth ceases with winter. Then, with the arrival of spring, the whole cycle is repeated. The juxtaposition of the narrow vessels laid down at the end of one year with the fat vessels of the spring that followed produces a conspicuous growth ring that can be seen in a cross section of the trunk. One ring marks the passage of a year in a tree's life.

You need only visit a botanical Shangri-La in the mountains of California to see signs of what ancient trees have had to endure in their long lives. Take, for example, the grove in Sequoia National Park where the giant redwood named General Sherman grows. The Sherman Grove is a place of both raucous resort and pious pilgrimage. Boisterous visitors vie with awed nature lovers, all jostling for a photo opportunity in front of the giant tree. There it stands, a 2,000-year-old behemoth weighing as much as six jumbo jets. Like a miracle, it once germinated from a seed the weight of a grain of rice.

General Sherman, shrewdly given the name of a military hero to protect it from loggers in the nineteenth century, stands like a giant fluted pillar from Zeus's own kingly temple. Scarcely tapering from its massive base, the column rises 200 feet to a blasted crown that is no longer capable of growing taller. The trunk is bare of limbs till nearly halfway up, but the aerial forest of branches that spring like independent scions from its side furnish enough nourishment to their native column that it grows in girth each year by a volume of wood big enough to make a good-sized oak tree.

What is the secret of the giant redwood's longevity? It is not the longest-living tree species by two or three millennia, but nothing lives 2,000 years without surviving events that kill most other organisms, and there are signs of commonplace mortality in the forest all around. The ground is littered with the fallen trunks of trees that died when only a few hundred years of age. Between the trunks is a drift of tiny plants that germinate, flower, set seed, and die all in less than a year. The black bear that rips with powerful claws at the rotten wood of the fallen trunks to get at beetle grubs is a large beast with no natural predators to fear, but he lives only 50 years if he's lucky. We humans in our natural state do a bit better. Hunter-gatherers can make it to 70.[5]

The redwood has its battle scars left by brushes with mortality. General Sherman and the equally ancient comrades in arms that surround him all have huge triangular gashes, ripped in their bark by fires. These charcoal-black wedges at the base of nearly every giant redwood are a story or two high and penetrate the thick bark right to the sapwood. The scars are testament not only to trial by fire, but also to victory over it. Redwoods have tough, fire-resistant bark.

Ancient bristlecone pines look even older, more weathered, and more severely tested by their long lives. Are such ancient trees senescent? There are too few of the trees to answer this question by using an increase in the mortality rate with age as a signature of senescence, as we did in shorter-lived species, but instead we can look for impairment in vital functions. Quite remarkably, a study of this kind that compared ancient bristlecone pines that were thousands of years old with youngsters of only a few decades found that growth of the cambium was just as vigorous in the ancients as in the young trees. So was the rate of growth of new shoots. Even the pollen and seeds produced by the ancient trees were as viable as those of the young ones.[6] The gnarled and twisted branches of ancient bristlecone pines give a deceptive impression of decrepitude, but due to recent warming of the climate near the tree line where they live, these ancient trees are now growing faster than they have at any time in the last 3,700 years.[7]

All the longest-lived trees are conifers, but there are only 627 species worldwide.[8] This low and rather precise number may creep up a bit with new discoveries such as the Wollemi pine, found as recently as 1994 growing in a canyon near Sydney, Australia, but it is probably as accurate as any species count can be. By contrast, the other main group of seed plants is the angiosperms, or flowering plants, and these are so numerous that their exact number is unknown and probably unknowable. According to one estimate there are about 300,000 angiosperm species, of which about 60,000 are trees, distributed among many families.[9]

The majority of angiosperm tree species live in tropical forests. If there is one adjective that seems to spring to the minds of authors describing a tropical forest, especially on their first encounter with it, it is "primeval." This description is a natural reaction to the darkness that prevails beneath a high, closed canopy and to the stature of the largest trees, but however ancient the tropical forest, how old are the trees themselves? Such a question is easily answered for temperate trees by counting annual growth rings, but in the tropics there is no cold winter to seasonally halt tree growth, so for a long time it was thought that tropical trees do not have annual growth rings. In fact, this assumption is simplistic, and though they may not show the very abrupt changes in growth seen in temperate trees, tropical trees do not grow at a uniform rate throughout the year. Most tropical environments have some seasonal variation in climate, though it is in rainfall rather than in temperature. These seasonal changes affect tree growth and leave behind telltale variation in the wood of tropical trees. By comparing growth variation in the wood with very accurate measurements of trunk growth, by repeated censuses of marked trees, and by the use of radiocarbon dating, many tropical trees have now been aged.[10] The results caused a terrific tiff in the tribe of tropical tree timekeepers.

In 1998 a study of trees felled in a logging concession in the Central Amazon estimated that the biggest of them, belonging to a species called *Cariniana micrantha*, in the Brazil nut family, was at least 1,400 years old.[11] The oldest trees had an average lifetime growth rate of only 1/32 inch a year in diameter, compared with six times this in trees that were just 200 years old. These results provoked a fuss when first published because tropical forests are known to be highly dynamic, with high rates of tree mortality from storms and other causes that lead to turnover of the entire forest every 400 years or so.[12] How could trees be so much older than the forest? Some very fast-growing tropical trees live for only decades, not centuries. However, the evidence does now point to there being some millenarian trees in the Amazon rain forest and elsewhere in the tropics,[13] and it also confirms that the oldest trees are the slowest-growing ones.[14] These large, old trees with dense wood seem to be able to survive the vicissitudes of climate that periodically fell the rest of the forest, though none of them, so far as we know, nearly matches the age of the oldest redwoods or bristlecone pines.

Do shorter-lived trees senesce? Surely trees that do not reach a great age must also have their lives curtailed somehow. Fortunately, we do have data on mortality rates in shorter-lived plants, including trees, and some, such as the Mexican *Astrocaryum* palm, certainly do succumb at a rate that increases with age.[15] I studied another clear example myself in the Adirondacks of upstate New York, where the lives of balsam firs on Whiteface Mountain come to a very abrupt end. As trees on the mountain approach 80 years of age, ice and fierce winter winds kill their foliage, and these tallest, oldest firs die en masse while still standing.[16] Although this is an extreme example with an obvious environmental cause of mortality, it is not unusual for conifers and other trees to lose branches with age and to acquire a thinning crown that cannot be hidden with a cone-over. Are such trees senescent, or, like that of bristlecones, is their decrepitude only skin-deep?

This question has been investigated experimentally by removing shoots from the tops of old trees and grafting them onto younger ones. The results, in both conifers and broad-leaved trees, show that old shoots grow with the same vigor as young ones when they are grafted onto young trees.[17] The conclusion is inescapable: whatever it is that limits the life span of a tree, it is not an age-related decline in the capacity of its cells to divide and to produce vigorous growth or viable offspring.

As in animals, the capacity for cell division in plants is a double-edged sword. On the one hand, it is essential for the renewal and repair that are required for long life, but, on the other, each new cell is a potential mutant. Plants have a virtually unlimited supply of dividing cells, which ought to enormously amplify the risk of mutation producing runaway cell division. Although mutation—as well as attack by certain bacteria, viruses, and insects—can produce tumors, plants seem to be almost immune to damaging levels of cancer. We have that on the good authority of James Joyce, who in *A Portrait of the Artist as a Young Man* quotes a nonsense poem[18] from a spelling book:

> Wolsey died in Leicester Abbey
> Where the abbots buried him.
> Canker is a disease of plants,
> Cancer one of animals.

But then if you've read James Joyce, you won't be surprised to find him discoursing on any subject. One reason plants are spared fatal cancers must be that plant cells are immobilized by a boxlike cell wall that prevents them spreading around the plant body in the way that animal cells are able to do. The phenomenon of metastasis that kills so many cancer patients cannot occur in plants. It has also been suggested that the division of a cell is more tightly controlled by the influence of neighboring cells in plants than in animals, which makes it much harder for a single mutant plant cell to multiply out of control.[19]

Mutations do occur in plant buds, but their effects remain localized, so occasionally a bud will produce a shoot and then a branch that is obviously different from the rest of the plant. In horticulture such wayward branches are known as "sports," and they can produce new plant varieties of great commercial value. Many traditional apple and flower varieties originated in this way.[20] Mutational variation of this kind is surprisingly rare, however, perhaps because mutant cells are usually replaced by wild-type ones within the tissue where they occur.[21]

It seems that so far as its cells are concerned, every tree is potentially as long-lived as bristlecone pine, perhaps even immortal. Why, then, do trees vary in how long they live? The knowledge that trees have very different life spans is even older than the biological cosmology encoded in the medieval Great Pavement of Westminster Abbey. A version of that formula for the age of the universe is found in a traditional Irish poem that is reputed to date from four centuries before the Great Pavement was made. That version begins, "A year for the stake. Three years for the field" and ends "Three lifetimes of the yew for the world from its beginning to its end."[22] A stake of willow can be grown in a year, but a yew grows so slowly that it is primeval in age. Thus, the ages of willow stake and yew tree bracket all other lifetimes. Yew is a European berry-bearing conifer with ancient mystical associations and is a favorite of poets. William Wordsworth wrote of a specimen in the English Lake District:

> There is a Yew-tree, pride of Lorton Vale,
> Which to this day stands single, in the midst
> Of its own darkness, as it stood of yore:
> . . .
> Of vast circumference and gloom profound
> This solitary Tree! a living thing
> Produced too slowly ever to decay;
> Of form and aspect too magnificent
> To be destroyed.[23]

Wordsworth's phrase "a living thing produced too slowly ever to decay" echoes, or rather anticipates, the association that we now know exists between slow growth and long life. In fact, not long after he wrote the poem, the Lorton yew was split by a storm and reduced from 27 feet around its middle to half that, but both halves had an afterlife. The half that broke off became a chair, supporting the posterior of the mayor of the nearby town of Cockermouth, Wordsworth's birthplace. The half that was left behind still survives in Lorton.[24]

As a tree grows in diameter, the xylem vessels become compressed to form the heartwood of the tree. These vessels no longer conduct water, but they help to give the trunk its mechanical strength. The physical and chemical characteristics of wood vary enormously among tree species and determine how well a tree resists attack by fungi and insects as well as physical damage from wind and the impacts of other trees as they fall. Wordsworth's rule of tree survival turns out to be a completely general one, as true in the tropics as in Lorton Vale. Trees with dense wood produced by slow growth have low mortality rates and long lives, while those that grow fast, like willow or birch, live only decades before they decay and die.[25]

Long-lived trees also defend themselves with chemicals. The fragrant resin produced by conifers, for example, is an important part of their armory, flooding wounds with antiseptic when the trees are damaged. The dried heartwood of a ponderosa pine can contain as much as 86 percent resin by weight.[26] Oil extracted from eastern red cedar wood is an effective termite and moth deterrent. Chests lined with the wood were traditionally used in New England to store and protect winter clothes from attack by moths during the summer months. Defensive compounds tend to darken the wood, so you can tell at a glance that the white wood that goes into plywood, for example, needs to be chemically treated if it is to resist rot. By contrast, the red aromatic wood of the western red cedar is naturally rot- and insect-resistant, even though it is also very light in weight. These properties make it ideal for use in outdoor construction. I have a greenhouse built of this marvelous wood and can testify to its rot-defying properties. The oldest western red cedar trees rival redwoods for their size and majesty and can live well beyond a thousand years. Perhaps it should not be surprising that chemically defended species live longer than their chemically defenseless relatives not only among plants, but also among fishes, amphibians, and reptiles.[27]

Even among members of the same species, the rule seems to be: Live fast, die young. There is no better illustration of this rule than the eastern white cedar, which, as we have seen, is a short-lived tree that dies within a century when able to grow fast in deep forest soil, but becomes a millenarian when forced to eke out a living in a rock crevice. Several studies of tree rings have found that the oldest surviving trees in a population are those that, compared with their fellows, have maintained relatively slow growth throughout their lives.[28] This finding is surprising because faster growth makes bigger plants, which one might imagine would survive better, but the cost of faster growth seems to be poorer resistance to stress. For example, experiments with several species of perennial herbs—including burdock, spear thistle, and foxglove—found that under normal conditions fast-growing individuals survived as well as did more slowly growing members of the same species and produced more seeds than they did. But when investigators stressed the plants by removing leaves, the subsequent survival and reproduction of the fast growers were much worse.[29] Slow-growing individuals stored away resources that faster-growing ones used in growth, and under stressful conditions the savers had an edge in recovery over the spenders.

Mortality in natural environments is often episodic. There are good times when few individuals of any age die, and there are bad times when mortality is high and hidden weaknesses are tested by adversity. This pattern is seen in laboratory studies of *Caenorhabditis elegans*, which have shown that stress favors wild-type over *daf*-2 mutants, even though the mutants live longer in unstressed conditions.[30] Similarly, a study of senescence in long-leaved plantain found that the expected rise in mortality rate with age was not detectable in protected greenhouse conditions, but was significant in a field population when a drought occurred.[31]

Large trees get all the press notices for extreme longevity, but there are plants that live even longer. Take an example I encountered on a field trip to South Africa. In England King Edward VII is a spud, but in the Diepwalle forest in the Cape region of South Africa he is a tree. Of kingly girth, he is 23 feet around the trunk, 130 feet high, and more than 650 years old. His thinning canopy of sparse branches towers over the rest of the forest, each branch tipped with a cluster of gray-green foliage and bearded with pendulous skeins of yellow-green lichen that hang in profusion beneath them. If ever there was an Ent of J. R. R. Tolkien's legend, it is this lofty, bearded tree. The guidebooks will tell you that this Outeniqua yellowwood[32] is one of the oldest trees in South Africa, but that is not so. King Edward VII misses that distinction by 10,000 years. At least he does if you believe the claims of a possible pretender to his crown.

Only 60 miles or so from Diepwalle, in the arid zone known as the Little Karoo, is a place called "Far-Away," or Vergelegen in the local Afrikaans language. I visited Vergelegen with Jan Vlok, a botanist who knows the Little Karoo like no one else. We set out from the town of Oudsthoorn, a sleepy, contented place in the middle of the Little Karoo. Jan folded his lean frame behind the wheel of the four-wheel-drive vehicle that South Africans call a "bakkie" and lit his customary cigarette, and with a puff of smoke and a touch on the gas pedal, we were off.

Twelve miles down a blacktop road, we turned off down a dirt track toward the mountains and sped, pursued by a tail of red dust, between wire fences that penned in stock with their diet of aridland shrubs. Suddenly Jan pulled the bakkie to a stop. We got out, and he gestured toward a small, unprepossessing tree just inside the fence.

"There's the gwarrie tree," he said.

"That's it?" I queried, the disappointment no doubt obvious in my voice. Jan had already told me that the oldest trees in the Little Karoo were small, but I was still quite unprepared for something that Jan said was at least 10,000 years old to look so ordinary. This is the story he told me about the gwarrie.

Long ago in Far-Away, at the end of the last ice age, some 10,000 to 12,000 years ago, the climate was wetter than it is today, and the gwarrie tree inhabited subtropical scrub. The whole gwarrie population in the Little Karoo must have been founded by the arrival of just one seed, or at most a very few seeds, because all the gwarrie trees alive today are genetically identical. Conditions must therefore have been very good, because from this modest beginning a whole population grew. Then the climate became drier, and successful establishment of new trees ceased. Nowadays, Jan said, the gwarrie's seeds occasionally get enough rain to sprout, but the seedlings invariably die of drought before they can grow a root long enough to reach the moisture buried deep in the arid soil that they need to sustain them. The only seedlings that survive are those that sprout in nearby plum orchards that are irrigated. Rainfall in this part of the Little Karoo is rarely more than 24 inches a year and often much less. Gwarrie seedlings need three or four years of good rain to become established, and the climate records for the area show that this never happens.

Could there really have been no new trees added to the population for 10,000 to 12,000 years? If so, then today's trees are at least that old, and the whole population is a museum piece stocked with exhibits twice as old as anything in the oldest Egyptian pyramid, and twice as alive too. Suddenly the gwarrie began to look more interesting. But how can gwarrie trees survive so long? The answer is hidden away belowground. Each tree sprouts from a subterranean bole of substantial size, a woody survival capsule that can resprout and replace the tree when it is burned or becomes an elephant's lunch. I was dying to see what lay beneath the gwarrie, but naturally Jan could not oblige without the aid of a gang of laborers, an obliging landowner, and official permission to interfere with a protected species.

The evidence for the record-breaking longevity of the gwarrie is, at least right now, circumstantial, but it is plausible because the gwarrie is not unique. The creosote bush is an aridland shrub that is native to the deserts of the American Southwest. It spreads by underground roots that radiate from the shrub and sprout to produce new, genetically identical bushes around it. Plants (and colonial animals such as corals) that are made up of numerous genetically identical individuals are said to be clonal.

As the older creosote bushes die, a circle of new ones is born that, through replacement, creeps outward from the original center like a ripple on a pond. This ripple travels extremely slowly. Taking modern rates of growth as a guide, the largest ring in the Mojave Desert, called "King Clone," has been estimated to be 11,700 years old.[33] This means that these creosote bush clones are as old as the Mojave itself, which, like the Little Karoo in South Africa, became a desert at the end of the last ice age.

Clonal plants can reach enormous ages,[34] but some biologists argue that old clones like the creosote bush, the gwarrie tree, and many others, which are comprised merely of young scions of an ancient genetic lineage, are in a different class from ancient bristlecone pines or the Outeniqua yellowwood and should not be regarded as old in the same sense at all.[35] These royalists would crown King Edward with glory when he is an old tree, but would cast out as a pretender a centenarian potato clone of the same name. In truth, the difference between the two is far less than it appears because all the really old parts of any ancient tree are dead. It is the young shoots of old trees that keep them alive. Arguably, the only real difference between an ancient bristlecone pine and a creosote bush or a potato is that the branches that connect the young shoots of a pine are aerial, while those that connect, or once connected, the young shoots of a creosote clone or potato are subterranean.

Lofty royalist or underground insurgent? I am a leveler by nature, but you decide. Biologically, however, all that the difference seems to come down to in the end is how long the connections between young shoots endure. If the connections are aboveground, as in trees, then the durability of wood, though dead, is crucial to longevity, because the shoots depend on the trunk for support and to provide a pathway to the roots. If the connections are underground, as in creosote bush, then the shoots can each develop an independent root system, and connections between bushes are not as important. Clonal plants differ a good deal in how long the connections last. In wild strawberry plants, for example, the connections are short-lived, while in bracken fern, which forms clones that are hundreds of years old,[36] they are longer lasting.

Whether long-lived clones senesce is an interesting question. Indeed, how would we know if they did? Measuring the mortality rates of old clones is next to impossible, but for those clones that reproduce sexually, we might look for a decline in sexual function as an indicator of senescence. A study of this kind in British Columbia, Canada, measured viable pollen production by trembling aspen trees belonging to clones that were up to 10,000 years old.[37] The age of an individual tree did not affect male fertility, but the age of the clone to which it belonged did. However, in 10,000 years, the loss of male fertility in old clones was only 8 percent. Though this was a statistically significant decline, habitats change so radically over such a timescale that this slight degree of senescence is probably of no biological importance. Compare it with the rapid loss of fertility in men, which declines by a third between the ages of 30 and 50.[38]

There is one group of plants for which senescence is predictable, abrupt, and terminal: the annuals. Some of them, like poppies, have spectacular blossoms, but others, like thale cress, have only tiny inconspicuous flowers that are the merest vestiges of ancestral sex. All germinate, set seed, and die within months. What causes their precipitate fall? The answer is surprisingly simple, once you understand how plants grow.

Plant growth originates from collections of cells that are dedicated to the job of dividing to make more cells. These founts that spill forth new cells are found in animals too: they are the stem cells that renew the lining of the gut twice a week and replace other cells around the body. In plants these founts are called meristems. The cambium is a layer of meristem cells dedicated to producing the specialized cells of the phloem and xylem. Every bud and growing tip of every branch has a meristem as well, which produces either a new shoot or a flower. A new shoot has its own meristem, so its growth can continue indefinitely, but flowers do not have meristems. So when a bud produces a flower, the stem or branch can no longer grow along the same axis.

An annual is a plant that, after a short life, switches all its available buds to flower production, thereby terminating vegetative growth. This burst of reproduction consumes all the plant's available resources, so any remaining buds that have not been switched to the flowering state lack the wherewithal to grow, and the plant dies. By contrast, a perennial survives from year to year by reserving some of its shoots for growth, allowing only a fraction of the buds that could produce flowers to actually do so. Flowering in perennials does not usually take place until the plant is large enough to bear the cost and survive, but annuals will usually flower when the right season comes around, however small the plant is. In many annuals even a diminutive plant a quarter of an inch high will sport a flower to its own funeral.

Flowering is usually triggered by seasonal environmental cues, but whether a plant responds to those cues, and by how much, is under genetic control. Thus there are flowering genes that ultimately determine whether a plant behaves as an annual and dies young or behaves as a perennial and delays senescence.[39] It is remarkable how similar the life of an annual plant is to the life of that short-lived animal *Caenorhabditis elegans*. Life span in both may be extended by throwing a genetic switch, but for some reason evolution prefers to snuff them both out in a paroxysm of reproduction.

Looking back over the variety of examples of longevity and senescence that have been discussed in this book so far, some patterns have become clear, and one big unresolved question stands out. What has become clear is that senescence, or the progressive deterioration of biological function with age, is one of the determinants of life span, but not the most important one. This is demonstrated in our own species by the extraordinary doubling of life expectancy that has taken place over the last two centuries. Senescence in our species has been progressively postponed, but not reduced.

Though most animal species seem to be subject to senescence, and some plants and modular animals apparently are not, this difference influences only the extremes of life span that are attainable by these two groups. The longest-lived modular organism has a life span measured in thousands (conifers and corals) or tens of thousands (clonal plants) of years, while the record holder among nonmodular organisms is a mollusk, the ocean quahog, which barely makes 500 years. However, most species of animals and plants have very much shorter lives. Short-lived plant species such as poppies die at the end of 12 months, and some relatively short-lived trees begin to senesce at around a century, but this is not caused by an intrinsic limit on plant cells' ability to divide and grow. Rather, it is a failure of body maintenance that is tolerated, perhaps even favored, by evolution.

We have discovered that evolution has the power to alter life span. This is obviously the case if you consider how much life expectancy varies between related species, from a year or two in mice to at least ten times that in the naked mole-rat, taking only rodents as an example. This variation among species demonstrates that longevity has a genetic basis, but what is more surprising is the existence of genetic variation for longevity within species. Analysis of the genes that cause this variation in nematode worms has revealed another surprise, in that essentially the same genes influence longevity from yeast to humans. The genes in question are involved in how organisms regulate the use of nutrients and how these nutrients are shared among the competing demands of growth, reproduction, and body maintenance.

Thus, in all organisms, including plants and corals, life span seems to be set by a flexible compromise among the options to grow, to reproduce, and to repair. This conclusion brings us to the big unresolved question that we need to address next: if senescence can be postponed, and if life span is so elastic, why does natural selection not expunge senescence and stretch life span indefinitely?

6
The Visionary Solution
NATURAL SELECTION

E, I sing for Evolution
V, the Visionary solution
O, the Origin of species
L, the Life that never ceases
V, the Victory for mankind
E, Emancipate the mind
The finger points in one direction
That's natural selection

STEVE KNIGHTLY, "EVOLUTION"[1]

There is a traditional tale of the Hausa, a people from West Africa,[2] in which two old men were traveling together on a long journey. They were hot and weary, their clothes ragged and full of dust, and the drinking gourds they carried were empty, so they decided that they must seek fresh water. They found a dry streambed and followed it until they eventually came to a spring bubbling from rocks at the foot of a hill. Next to the spring there was a young man sitting on a rock, so they asked his permission to drink.

"By all means," he replied. "But let the elder of you drink first, for that is the custom here."

"I am Life," said one of the old men, "and I am the elder."

"No you are not," replied the other, "for I am Death and I am the elder."

"That cannot be," said Life, "for without Life there cannot be Death, and so I am older than you."

"On the contrary," said Death. "What existed before life was born? Only nothingness and death. I am much, much older than you."

The youth by the spring could see that this argument wasn't going to be settled quickly, but out of respect for Life and Death he sat patiently on his rock and waited for one of the old men's thirst to overcome his pride. Eventually, Life turned to him and said, "All right, young man, you have heard what we have both had to say. You choose who is the elder, Death or me."

The young man was worried by this request because he was afraid that to favor one old man would upset Life, while to favor the other would provoke Death. So he diplomatically replied:

"I have listened to all that has been said, and both of you are wise and have spoken the truth. There cannot be life without death or death without life, and so you are equal in age. Neither is the elder. You must both drink." So saying, he handed the two old men a large bowl of clear spring water, from which they drank eagerly together.

This view of death prevails, not just among the Hausa, but everywhere. Life and death are inveterate traveling companions that drink from the same cup. In youth Life races ahead, oblivious of his mortal shadow, but as he grows older, that shadow comes closer and closer until Death catches up. This is the universal human experience. Many writers have sat by the same metaphorical spring, witnesses to the contest between Life and Death, and offered their own judgment. American poet Emily Dickinson (1830–1886) wrote:[3]

Death is a Dialogue between
The Spirit and the Dust.
"Dissolve" says Death—The Spirit "Sir
I have another Trust"—

In a similarly religious vein, the sixteenth-century English poet John Donne (1572–1631) proclaimed, "Death be not proud,"[4] for there is an afterlife:

One short sleep past, we wake eternally
And death shall be no more; Death, thou shalt die.

Equally inspired by scripture and by a bet with another poet as to which of them could write the better poem on immortality,[5] Welsh poet Dylan Thomas (1914–1953) wrote of death as a liberation from mere mortality:

And death shall have no dominion.
Dead men naked they shall be one
With the man in the wind and the west moon;
When their bones are picked clean and the clean bones gone,
They shall have stars at elbow and foot;
Though they go mad they shall be sane,
Though they sink through the sea they shall rise again;
Though lovers be lost love shall not;
And death shall have no dominion.

The same idea, that death will be vanquished by dying, was expressed by the Roman poet Seneca (ca. 4 BC–AD 65),[6] though he firmly rejected the idea of an afterlife:

After death nothing is, and nothing, death:
The utmost limits of a gasp of breath.
Let the ambitious zealot lay aside
His hopes of heaven, whose faith is but his pride

Seneca's view is nearest to that of modern science: death is the end of life and nothing more. However, scientific curiosity does make us ask, "Why?" Why must Death always catch up with Life? After all, there are species that are so long-lived that they seem to be virtually immortal. These are mainly plants, to be sure, but even among animals, Death holds Life on a much longer leash in some species than others. And in our own species, we have stretched that leash by 15 minutes every hour. The evidence therefore tells us that length of life is a changeable thing and that the timing of death, like everything else in life, can be altered by evolution. Therein lies a puzzle.

Natural selection, the motor of evolution, favors the individuals that leave the most descendants, so how can aging, which impairs fertility and causes degeneration of the body, evolve? Why does natural selection permit aging? Why hasn't natural selection fixed the problem and made individuals of all species immortal? One of the first scientists to ask this question was the nineteenth-century German biologist August Weismann (1834–1914). He suggested that aging and death are favored by evolution because they benefit the species by removing worn-out individuals, making way for the young and vigorous.[7] Unfortunately, this superficially attractive idea is thrice flawed, as Weismann himself eventually realized.

First is the problem that natural selection does not work for the good of the species, but instead works on individuals, favoring those whose inherited traits cause them to leave the most descendants. Natural selection for individual advantage trumps any alternative that would involve a sacrifice purely for the good of the species. To see why this is so, imagine a population in which old individuals sacrifice themselves, as Weismann envisaged, for the good of the species. Sooner or later a mutant would appear with a defective gene for self-sacrifice. By living longer, this mutant would be able to leave more offspring than any self-sacrificing individual, and in just a few generations self-sacrifice would go distinctly out of fashion.

Second, there is a problem with the notion that organisms wear out as if they were machines. Why should biological processes that can perform amazing feats of development like turning an egg into a chicken have difficulty repairing the chicken once grown? Hence senescence cannot simply be a matter of organisms wearing out from absence of repair, though it might happen through neglect. So if organisms do appear to wear out, this does not explain senescence at all, but rather changes the question to: Why would an old organism fail to repair itself when a younger one is capable of doing so?

This question reveals the third and final problem with Weismann's theory: that it is circular. It does not explain how senescence would evolve from a starting point where it does not already exist. Instead, the theory assumes that senescence exists. Weismann argued that removing individuals worn out by age would benefit the species, but this does not explain why individuals wear out with age in the first place. So we are back to the original question: Why does natural selection tolerate senescence?

The first person to come up with a convincing and explicit evolutionary explanation was British biologist Peter Medawar (1915–1987), who tackled the subject in an article in an obscure magazine in 1946 and then reprised his ideas more fully in a published lecture called *An Unsolved Problem in Biology*, published in 1952.[8] Perhaps Medawar's discovery would have attracted more attention at the time if he had titled his lecture *A Problem in Biology Solved*, but as he says in his biography, *Memoirs of a Thinking Radish*, he was only dabbling in evolution for intellectual amusement. His day job, so to speak, was as an immunologist, and his discoveries in that field won him a Nobel Prize in 1960. A second Nobel for solving the evolutionary problem of why we age would have been well deserved. I came across Peter Medawar myself, though only from the remote distance of the back of a lecture hall, in 1979. By that time he was in a wheelchair, incapacitated by a stroke. He had tragically become an illustration of his own argument for how senescence evolves.

Medawar's argument is elegantly simple, and unlike Weismann's theory, it is completely consistent with natural selection. Imagine a population in which there is no senescence, so that the mortality rate does not increase with age, and in which death is entirely due to random accidents. If the birth rate and death rate are both constant over time, such a population will arrive at an age composition that is dominated by the young. Accidental death alone will ensure that the number of survivors will diminish with advancing age. The progressive thinning of the ranks of the old occurs simply because the longer you live, the more chances there are that you will meet with a fatal accident. Now imagine that nearly every individual, from adolescence to old age, can have children. Fast-forward by a generation and ask everybody in that generation how old their parents were when they (the children) were born. The average age of parents will be young, just because most individuals in the population are young.

Medawar's great insight—helped, it must be said, by some strong hints from another genius, J. B. S. Haldane (1892–1964)—was that the situation described would lead to the accumulation of deleterious mutations that act late in life. This could happen because such mutations would be passed on to children before the effects struck their parents. By contrast, early-acting mutations would be much more likely to impair reproduction in parents and would consequently limit their own transmission to future generations.

A clear example of a late-acting mutation is the single faulty gene that causes Huntington's disease, the neurodegenerative effects of which do not emerge till patients are in their fifties. The

American folksinger and political activist Woody Guthrie (1912–1967) inherited Huntington's from his mother, but by the time his symptoms became disabling, he had fathered at least seven children. While Peter Medawar's own ill health may or may not have had a genetic basis, he too had completed his family of four children before his first stroke.

More common neurodegenerative diseases such as Parkinson's and Alzheimer's, as well as other diseases such as stroke, cardiovascular disease, diabetes, and cancer, all occur mainly in later life too. The role of inherited mutations in these diseases is much less clear-cut than in the case of Huntington's, but where inheritance plays even a small role—for example, through the influence of the *APOE* gene mentioned in chapter 4—the mutations involved will accumulate beyond the reach of natural selection.

There is a modern trend toward people beginning their families later in life. This delay may cause natural selection to begin to act against deleterious alleles, such as the *APOE* ε4 allele, that previously did their damage too late in life to affect reproduction. Thus one might expect the frequency of the ε4 allele to begin to diminish as it is increasingly caught in the searchlight of natural selection as its sweep advances deeper into our lengthening reproductive life span.[9]

In summary, Medawar's idea is that the ability of natural selection to alter the genetic future diminishes with the age of individuals and that this, by default, permits mutations that cause senescence to accumulate over evolutionary time. One might say that natural selection retires in old age.

Peter Medawar went a step further with his argument, pointing out that some mutations that have beneficial effects on health and reproduction during youth might also have deleterious effects in old age. Such double-acting mutations would accelerate the evolution of senescence because they would actually be favored by natural selection and not just passively accumulate. Double-acting genes that have reproductive benefits in youth but health penalties in old age can be compared to a children's seesaw, with life span represented by a plank that connects youth and old age. Raising one end of the seesaw results in lowering the other. Natural selection elevates youth, but it is indifferent to the plunge in old age that results at the other end of the plank.

A major group of the diseases that afflict humans in old age are related to the immune system.[10] In youth, a well-functioning immune system protects us from infection and has obvious survival value. Vaccination, which has dramatically cut child mortality in the last 100 years and has thus increased life expectancy, works by readying the immune system to fight specific viruses and bacteria before they strike. But in old age, the immune system can become oversensitive and prone to cause inflammation in the joints, resulting in rheumatoid arthritis.

There is genetic evidence that mutations that increase susceptibility to rheumatoid arthritis have actually been favored by natural selection during our recent evolutionary history.[11] This finding suggests very strongly that the mutations in question are double-acting and must have beneficial effects during youth. It is not known how recently the selection started, but it may have been triggered by the advent of agriculture about 10,000 years ago, which exposed humans to many new diseases.[12] Agriculture also greatly increased the density of human settlements, making the transmission of disease much easier. Such conditions would have produced strong selection for mutations that improved the response of the immune system to disease, regardless of any consequences in old age.

American biologist George C. Williams (1926–2010) followed up on the idea of mutations that have beneficial effects early in life and deleterious ones later on by deducing a series of important predictions that are made by this version of Medawar's evolutionary theory of senescence.[13] In fact, the same predictions apply equally to Medawar's simpler mutation accumulation theory. First was the prediction that for senescence to evolve, there must be a separation during embryonic development between germ line and soma. The "germ line" sounds like an unsanitary branch of the New York subway system, but it's actually the lineage of cells in the body that produces eggs and sperm. The soma (which means "body" in Greek) is the rest of the organism. Because the germ line is the route by which genes are transmitted to future generations, a mutation that impairs germ cells in any way will spell its own doom. This is not so for mutations that damage somatic cells if they act only after reproduction has taken place. Hence natural selection will tolerate the deleterious effects of mutations that cause senescence only if the germ line is protected from their effects. Note that these mutations are transmitted in the cells of the germ line, but they manifest their effects only in the soma.

The separation of germ line and soma is normal in most animals, so Medawar's theory predicts that senescence can evolve in those animals. However, there is no separation between germ line and soma in plants. The ovules and pollen grains in a flower and the cells that make leaves and the branch to which they are attached all have a common origin in the handful of meristem cells that made the bud from which they sprouted. Hence, Williams argued, mutations that cause senescence cannot be favored by natural selection in plants. This argument would account for the huge ages that plants, and some plantlike animals such as corals, can reach (described in chapter 5). There are plants, such as annuals, that clearly do senesce, but their life cycle must have evolved by a mechanism not involving double-acting mutations or mutation accumulation. We'll take a look at some spectacular examples in the next chapter.

Another condition necessary for mutations to cause the evolution of senescence is that the number of offspring produced must diminish as an organism gets older. This is the situation in humans and in the domesticated animals, like dogs and livestock, that we are most familiar with, so it may sound normal, but there are perhaps 10 million species on the planet, every one of them a bit different, so we should be cautious about calling anything "normal." Animals and plants that have indeterminate growth, and are therefore capable of getting bigger and bigger with age, violate this particular condition. In these species the seesaw is too finely balanced by the numerous offspring of old parents to permit natural selection to sacrifice them in favor of youthful advantage. Indeterminate growth, and the resulting pattern of increasing fecundity with age, may well be why long-lived trees and, among animals, bivalves like the ocean quahog live to such a great age.

The evolution of senescence demonstrates that natural selection ultimately cares only about reproductive success. This conclusion highlights another evolutionary puzzle: Why does fertility in women cease at 50? Menopause occurs at around this age in all human populations. Men's fertility also declines with age, but it is not abruptly terminated, as it is in women. To add to the puzzle, menopause does not occur in any of our primate relatives; female chimpanzees, for example, reproduce until the end of their lives. So menopause, unlike senescence, seems to be peculiar to humans. It is possible to reverse menopause to some degree with hormone treatment. In India in 2008 a Mrs. Rajo Devi gave birth at the age of 70 after IVF treatment.[14] All these facts suggest very strongly that menopause is not a mere by-product of natural selection or an inevitable consequence of senescence, but, paradoxically, must have evolved because it somehow confers reproductive advantage.

Since menopause halts reproduction, it can confer reproductive advantage and aid the transmission of the genes involved only if it benefits a woman's existing children or grandchildren. Furthermore, the size of this benefit must be larger than the cost to a woman's reproductive success of ceasing to produce more children of her own. In other words, when natural selection does the sums, the net benefit, calculated in numbers of descendants, must lie with menopause, not with continuing reproduction. Two factors influence this calculation: first, how many babies of her own a woman could expect to raise successfully after the age of 50, and, second, what difference she could make to the survival and reproduction of her existing children if she helped them instead.

The answers to these questions clearly depend on prevailing health and social conditions, all of which have improved in recent times, but remembering this caution, we can still try to estimate how the numbers might have stacked up in our evolutionary past. Most women have had the majority of their children well before 50, and there are dangers in having more at that age. The risk of maternal death in childbirth increases with a woman's age, and so does the risk of conditions such as Down's syndrome in her children.

Data on such numbers are hard to come by further back than about 150 years, but a remarkable study utilized the enduring British obsession for the weddings and beddings of its aristocracy to trace their record of reproduction back 1,200 years.[15] Among both ladies and lords, eminent personages with larger families had shorter lives, particularly in the premodern period before AD 1700. Nearly half the women who lived to 81 years of age had borne no children at all. Even when death in childbirth is disregarded, as it must be for fathers, these data and other studies show that having children must have carried a cost in length of life during most of human history. If this cost affected aristocrats, it most certainly affected peasants, who lived a much harder life.

Such data suggest that the risks of having further children after the age of 50 might easily be outweighed by the benefits of not doing so. By that age a woman's oldest daughters would probably have their own babies, and helping them raise those babies could increase the number of her grandchildren that would survive, and perhaps the number that would be born. This "grandmother hypothesis" for the evolution of menopause is supported by some evidence. A study of two villages in Gambia, West Africa, using data collected before medical facilities became available there, found that children one to two years of age who had a maternal grandmother in the family had twice the chance of survival of children the same age who had no surviving maternal grandmother.[16] In another study that used church records of births and deaths in premodern Finland, grandmothers who survived beyond 50 had more grandchildren than those who did not.[17] In neither Gambia nor Finland did the survival of grandfathers influence the survival or number of their grandchildren.[18] Perhaps this accounts for why women live longer than men, since grandfathers are apparently redundant in the brutal calculus of natural selection. There is a conspicuous dearth of men among the old.

Though menopause occurs in no other primate, it is not totally unique to humans. It occurs in another group of mammals: toothed whales. Female orcas (killer whales) cease to reproduce at around 40 years of age, though they may live into their nineties. Male orcas, like male humans, are able to reproduce throughout life, though, also like humans, they do not live as long as females. A remarkable study of orcas living off the coast of the northwestern United States and Canada found that these animals, which live all their lives in permanent family groups called pods, survive much better, even as adults, if their mother is alive than if she dies. This effect was especially strong for sons, who, even at the age of 30 or more, experienced a fourteenfold increase in mortality in the year after their mother died compared with males whose mothers survived.[19] How mothers help their adult sons to survive is not known, but future research into the behavior of orcas may reveal the answer.

What do humans and orcas have in common that might have caused the independent evolution of menopause in two such different mammal species? Two shared characteristics seem to be important in creating the conditions required for a post-reproductive female to be able to increase the reproductive success of her offspring, either as a grandmother (in the case of humans) or as a mother (in orcas). The first is that humans and whales are both long-lived. Only in an exceptionally long-lived animal is it possible for a female to live long enough to assist the reproductive success of later generations.

The second shared characteristic is that both humans and orcas live in family groups containing several generations—an orca pod may contain as many as five. Human families and orca pods create the social conditions in which an individual who assists those younger than herself is also indirectly assisting the transmission of her own genes to future generations through her relatives. Without such a close-knit family structure, natural selection would not favor a female who helped others at the expense of reproducing herself, and menopause would not evolve.[20]

The perennial battle of the sexes has produced a fog of prejudice and humor about how women and men suffer ill health. A common joke—perhaps it is even a common perception—is that men turn the slightest ailment into an attention-seeking drama. Surveys of ill health in women and men tell a quite different and paradoxical story. For the top twelve major health conditions, such as cancer and cardiovascular disease, mortality rates are higher among men than women at all ages. However, women have higher rates than men of physical illness, visits to the doctor, and stays in the hospital. When asked about their health in surveys, men report better health than women of the same age, but rates of mortality show that women are the more robust sex.[21] Women and men age at the same rate, but the baseline or initial mortality rate is lower in women than in men. If, in typical male fashion, I wanted to overdramatize the situation, I would say, "Women are born to suffer, men are born to die."

Although menopause appears to be unique as an evolutionary phenomenon, the underlying processes that drive it are not. Foremost among these processes is the trade-off between reproduction and survival. This trade-off is not a unique privilege of the British aristocracy; in fact, it is found in yeast, plants, worms, fruit flies, viruses,[22] and in practically every species at which anyone has ever looked.[23] Cast the net in the wider world, and trade-offs of one kind or another will be found everywhere from gluttony—"you cannot have your cake and eat it too"—to music. The modernist composer Arnold Schoenberg summarized his art as requiring a balance between "the demand for repetition of pleasant stimuli, and the opposing desire for variety, for change."[24]

In the scales of natural selection, the balance between survival and reproduction is weighed in units that measure contribution to future generations, referred to as "fitness." Evolutionary or Darwinian fitness is not to be confused with the kind you acquire by regular attendance at the gym. This point was eloquently made to his students by the evolutionary biologist John Maynard Smith, whose eyesight was so bad that he wore glasses with pebble lenses. Due to his poor eyesight, he was declared unfit for military service in the Second World War, which, he joked, probably saved his life and thereby raised his Darwinian fitness.

Trade-offs exact a price for longer life in mutant *Caenorhabditis elegans* that is paid in lower fitness. In experimental mixtures of *daf*-2 mutants and wild-type worms, the longer-lived mutants disappeared after only three generations because they produced fewer eggs in early life than wild-type worms did.[25] Another *C. elegans* longevity gene, called *clk-1*, is handicapped in a similar way.[26] These results are examples of the premium in fitness that is gained by reproducing early (see chapter 2). Resveratrol is a plant compound that has been credited as the source of the beneficial effects of moderate consumption of red wine. *C. elegans* that were fed with resveratrol lived longer, but they too produced fewer eggs in early life.[27] Whether this effect should worry red wine drinkers more than the well-known health risks of alcohol consumption is doubtful.

John Maynard Smith (he of the pebble glasses) discovered over 50 years ago that mutant fruit flies lacking ovaries lived significantly longer than wild-type flies, demonstrating that there is a cost of reproduction that is paid in length of life.[28] Later experiments with fruit flies and *C. elegans* suggested that reproductive cells generate chemical signals that flip genetic switches in the molecular pathways that control length of life.[29] Trade-offs, then, just like length of life and menopause, are under genetic control, but the ultimate arbiter of how the switches should be set is their effect on fitness. This effect, in turn, often depends on the environment. A *daf-2* mutant worm might look like a winner in a petri dish, but loses out badly to the wild type in the natural environment of the soil.[30] A British aristocrat living in the Middle Ages might pay the price of many children and die early, while in the improved conditions of the nineteenth century Queen Victoria could have nine children and live to 81. Interestingly, animals in zoos are cosseted like royalty, and in such favorable conditions they do not show the detrimental effect of reproduction on female longevity that is seen in wild populations.[31] It shouldn't be a surprise that the environment is so important because, after all, Darwinian fitness is highest when an organism is adapted to its environment. Such adaptation can favor surprisingly odd behavior, including suicidal reproduction, as we shall see in the next chapter.

7
Semele's Sacrifice
SUICIDE

JUNO: Above measure is the pleasure
Which my revenge supplies
Love's a bubble gained with trouble
And in possessing dies

WILLIAM CONGREVE, *SEMELE*, ACT 2

Though he died in another land and a thousand years before Westminster Abbey was built, the spirit of the Roman poet Ovid (43 BC–AD 17) haunts the place in the collective legacies of the poets interred and remembered there. The work that immortalized Ovid's name was his long poem *Metamorphoses*. This poem begins with the creation of the universe and ends in Ovid's own time, so its central theme of recurring transformation is a kind of evolutionary history of nature, albeit a mythological one. In an epilogue that brazenly challenges the gods and the emperor Augustus, who banished him from Rome, Ovid says that now that his poem is complete, nothing can destroy his work, not even Jupiter's wrath, and that he is ready for death whenever it should come because he knows that *Metamorphoses* will "sweep me into eternity, higher than all the stars. My name shall not be forgotten." He adds, "Throughout all ages, if poets have vision to prophesy truth, I shall live in my fame."[1] And Ovid was right.

From Chaucer's *Canterbury Tales* to Shakespeare's *Tempest* and Mary Shelley's *Frankenstein*, English literature is suffused with the influence of Ovid's *Metamorphoses*, in which he recounts the often vengeful transformations visited on mortals by Greek gods whom they had displeased.[2] Self-obsessed Narcissus rejected the love of the goddess Echo and was transformed into a flower. Actaeon was out hunting one day when he chanced upon the goddess of the hunt, Diana, bathing naked in a forest pool. To ensure that Actaeon could not tell anyone what he had seen, Diana transformed him into a stag, whereupon he was torn to pieces by his own hunting dogs. A happier transformation was granted to the sculptor Pygmalion, who fell in love with his own creation—a woman carved in ivory—when Venus, goddess of love, brought it to life in response to Pygmalion's offerings at her altar. Perhaps Henrietta, Duchess of Marlborough, had this transformation in mind when she commissioned the mechanical ivory statue of her lover William Congreve with which she habitually conversed after his death. Among the works of this prolific author is a verse translation of Ovid's story of the fate of Semele, which the composer Handel used as a libretto for a secular oratorio (in effect, an opera).[3]

Semele (pronounced to rhyme with "Emily") was the daughter of Cadmus, founder and king of the Greek city of Thebes. She is portrayed on ancient Greek vases, but Ovid's is the oldest written version of her story to survive. In Greek myth the Thebans are always attracting the attention of the gods, in particular Jupiter (aka Jove), who is partial to the women of that place. Jupiter has form: he has seduced Semele's aunt and raped an ancestress. So when Semele begins to show an unhealthy interest in worshipping at Jupiter's altar, the god's wife Juno takes exception. Sure enough, Jupiter whisks Semele up into heaven, and at this point in Handel's opera she can be heard rejoicing from cloud nine in a lyrical aria:

> Endless pleasure, endless love
> Semele enjoys above!
> On her bosom Jove reclining,
> Useless now his thunder lies;
> To her arms his bolts resigning,
> And his lightning to her eyes

Juno knows that Jupiter himself is incorrigible, so she decides instead to take her revenge on Semele, who is pregnant with her husband's child. She visits Semele in the guise of an old crone and asks Semele how she knows that her lover is indeed who he says he is. Men will tell you anything to get inside your toga. What you must do, says Juno, is make him promise to reveal himself in his true form, as he does when he sleeps with his wife Juno. You are entitled to no less than she is, surely? To ensure that Jupiter will not refuse her request when he hears it, Semele asks Jupiter to promise that he will grant her anything she wants. Totally smitten with Semele, Jupiter agrees, but when he hears her wish, he realizes too late that she has unwittingly asked for death, because in his true form he is a thunderbolt. Bound by Semele's ineluctable, fatal wish, Jupiter appears to her as his real self, and Semele dies. Juno has had her revenge, but that is not quite the end of the story. From Semele's ashes, Jupiter rescues their unborn child, which he sews into his thigh, where it completes its gestation. The baby is Bacchus, god of wine and merriment, and before the curtain falls, William Congreve sends his opera audience away on a high with these words from the chorus:

> Happy, happy shall we be,
> Free from care, from sorrow free;

> Guiltless pleasures we'll enjoy,
> Virtuous love will never cloy;
> All that's good and just we'll prove,
> And Bacchus crown the joys of love!

As you have probably guessed, there is a biological point to this story too. Babies may be a cause for joy, particularly if they arrive, like Bacchus, carrying the keys to the cocktail cabinet, but sex is hazardous. The mythical Semele is the ultimate example of an iron law that governs life: reproduction bears a cost. It is rarely as severe as the price paid by Semele for a single baby, but there are plenty of examples in nature in which reproduction is the herald of death. Biologists call this pattern "semelparity" after Semele herself.

Semelparity, or big-bang reproduction if you prefer, is scattered across the plant and animal kingdoms. To see a spectacular plant example, accompany me now to Kirstenbosch National Botanical Garden in Cape Town, South Africa. Sit here on this bench that bears a plaque dedicated to a deceased visitor to the garden, one Dieter Kern, who died in 1995 at the age of 51. I am already five years older than this, so this bench is a personal reminder of my mortality. Indeed, so wonderful are the flowers in Kirstenbosch, so stunning is its location at the foot of Table Mountain, and so celestial is the chuckle of a stream running fresh from its flanks and the chirrup of frogs calling to one another, that we might already be in heaven. But that cannot be, because any semblance of eternity about this place is contradicted by the trees standing right in front of us. They are a trio of kosi palms, native to the coast of Maputoland, in the east of South Africa. Two of the three trees are clearly dead, with leaves as brown as their trunks. For a memorial, these trees have left us a fibrous wood not durable or strong enough for even the most flimsy garden bench, plus a hill of fruits, most still attached to dead candelabra of branches in their crowns.

The fruits are the tree's promise of immortality. The trunk is merely an ephemeral means to that end, and it only needs to be strong enough to provide a temporary structure during life. In palms the trunk is in truth little more than an accumulation of leaf bases and does not grow fatter with age, as in broad-leaved trees (and broad-bellied humans). Each fruit is the size of a large hen's egg and is armored with scales of a varnished mahogany brown, arranged in a spiral pattern like the scales of an unopened pinecone. Fibonacci is the name of the mathematician immortalized by his discovery that spiral patterns like this conform to a series in which each number is the sum of the previous two: 1, 1, 2, 3, 5, 8, and so on. The Fibonacci series turns up time and time again in nature when she spirals.

The middle kosi palm of the trio has not fruited yet and is still growing vigorously, sprouting leaves like a fountain jet from its crown. These hundred-foot-tall trees produce massive 33-foot-long leaves that sprout one after another in spiral fashion around the trunk. As it grows, the kosi palm corkscrews itself skyward, producing a canopy of large feathery leaves sprouting from the crown of a slender stem, resembling a giant feather duster poised to spring-clean an untidy, cloud-strewn sky. At 30 years of age or more, the topmost bud of Atlas's feather duster switches from leaf production to producing a large candelabrum of flowers, which then bear the fruits that pretend to be cones. A true pinecone contains dozens of seeds, but a fallen kosi fruit rattles when you shake it, and inside its cavity is just one ivory-like nut. The woody scales protect the fruit from predators while it is ripening on the tree, but loosen and fall away to release the nut when it is on the ground and has to germinate. The kosi and many other palm species pay the ultimate price for producing a large crop of sizable seeds by dying in the effort. The talipot palm of southern India and Sri Lanka is the most spectacular example, with leaves even bigger than the kosi's and a gigantic candelabrum of fruit 10 or 12 feet tall crowning the tree at the end of its long life.

Semelparity is common in a few groups of animals and plants but very rare in others. Among plants, palms are the only trees that go in for it in a big way, though there is a handful of other tropical trees that are semelparous. Many species of bamboo are semelparous, and these species may flower simultaneously and then die across vast areas. In North America and Europe, semelparous herbs such as wild carrot, mullein, and evening primrose tend to live in weedy habitats, where they colonize gaps in the vegetation.

Once-only reproduction is very common in insects, some of which spend several years hidden away as aquatic or subterranean larvae before their brief moment in the sun. The larvae of dragonflies, for example, live in fresh water, where they are aggressive predators on other animals, including small fish. Periodical cicadas spend up to 17 years underground as nymphs, feeding on the sap of tree roots, before they simultaneously emerge as adults in huge numbers to mate, lay their eggs, and die. Pacific salmon make a one-way journey to their spawning grounds in the headwaters of North American rivers after three years of celibacy at sea. Eels make a one-way trip in the reverse direction, spending their middle years in fresh water and then converging from Europe and North America on the Sargasso Sea, where the European and American species both breed.[4] Many squid and octopus species are semelparous too, which is an important consideration in managing these fisheries because most of the individuals caught will not yet have reproduced.[5]

Semelparity is rare, but not unknown, among mammals. Most examples belong to a group of Australian carnivorous marsupials in which only the males die after a very intense and promiscuous bout of mating. Some snakes are semelparous, and a recently discovered tiny chameleon from Madagascar, called *Furcifer labordi*, spends most of its short life in the egg, living for only four or five months after hatching.[6] This chameleon is the only known vertebrate example of the annual life history that is so common among plants. Annual plants, as we saw in chapter 5, may spend decades as seeds in the soil before they emerge, grow, flower, and die within a brief season of a few months.

Semelparity is a fascinating phenomenon because concentrating all reproduction into a single bout incurs the ultimate cost of reproduction that any organism can pay. Why does the miscellaneous collection of species that we have just described share such an extreme and seemingly risky way of life? Could there be a single explanation for all cases, from palms to periodical cicadas and from bamboos to squid? Well, yes, there could.

Let's do another thought experiment and see how semelparity checks out against what we humans would consider to be a more normal way to raise a family. We'll start with an annual plant and assume that it produces 10 seeds at the end of the year, at which point it dies. The next year, all 10 seeds germinate, survive, and make their own 10 seeds apiece. Thus, after two years, the original semelparous annual has $10 \times 10 = 100$ descendants. Can this number be beaten by a mutant that refuses to die after setting seed? Survival requires some resources to be kept in reserve, so the mutant cannot produce as many seeds as the annual. Let's say that it parsimoniously produces 9 seeds, which economy spares just enough of its reserves for it to limp from winter into spring, after which it is able to produce another 9 seeds. At the end of two years, the first 9 seeds have each produced 9 seeds of their own, making $9 \times 9 = 81$. Add the 9 surviving plants, and the total is 90. Then add 9 seeds from the mutant's second year of reproduction and the original plant itself, and we get $90 + 9 + 1 = 100$ descendants. Not very impressive, is it?

The point of this labored piece of elementary arithmetic is that semelparity is harder to beat than you might have imagined. The annual has only to squeeze out one more seed (i.e., 11), and it can beat the competition by more than 20 percent ($11 \times 11 = 121$). American biologist Lamont C. Cole noticed in 1954 that by such a calculation one comes to the paradoxical conclusion that all species should be semelparous annuals, but of course they aren't.[7] Why not?

If you are still with me, you are no doubt looking for the catch, or thinking, "Ah, but what if . . . ?" And that is the real point of Cole's paradox. It immediately gets us asking questions: But what if not all the seeds survive? What if the adult plant survives winter only half the time? What if the adult can grow bigger in its second year? Like Peto's paradoxically cancer-free whales in chapter 2 and the paradoxes of natural selection tolerating senescence and favoring menopause in chapter 6, this paradox focuses our attention on an evolutionary puzzle that needs to be solved.

The mathematical solution to Cole's paradox turns out to be a remarkably simple rule, although the biological solutions that natural selection has engineered that fit the rule are wonderfully various and endlessly surprising. The rule is that for a repeat breeder to beat semelparity, the ratio of the number of offspring produced by repeat breeding to the number produced by semelparous individuals, added to the probability of survival of the repeat-breeding parent after it has reproduced, must be greater than one.[8] According to this rule, the simplest way for semelparity to be beaten is for the parent to always survive reproduction (i.e., survival probability = 1). But as we have seen, the biological reality is that reproduction always exacts a cost, and this cost often causes parental mortality. On the other side, the way for semelparity to tip the balance is to produce a sufficient excess of offspring over the repeat-breeding competition to compensate for the normal probability of parental survival. The lower normal parental survival is, the easier it is for semelparity to evolve. Now, that's more than enough math! How this rule translates into real lives is the cool stuff.

Tasmanian devils, the poor devils we met in chapter 2 that are afflicted with a newly emerged infectious facial tumor disease, provide a gruesome example of how high adult mortality can favor the evolution of semelparity. Before the disease appeared, Tasmanian devils reproduced over their whole life span once they had reached sexual maturity. However, adult mortality in infected populations is almost 100 percent in the second year of life, and animals now reproduce precociously, but only once before they die.[9] The rapid evolution of semelparity in these populations is dramatic confirmation of the influence of adult mortality as predicted by the solution to Cole's paradox. It also demonstrates how, when an animal adapts to new circumstances, evolution may help save it from extinction, though whether this species will actually survive in the wild remains to be seen.

Australia is also home to two families of small marsupial mammals in which males are semelparous but, curiously enough, females are not. In the brown antechinus, for example, females all come into season at the same time, and each female mates with many males, producing litters of up to eight young by as many as four different fathers.[10] This mating system produces continuous and intense competition among males for mates. Males' physiology overdoses on testosterone, and the bloodstream is flooded with corticosteroid stress hormones, causing body maintenance to be sacrificed to the effort to mate.[11] Males lose weight, lose fur, have a weakened immune system, and become parasite-ridden, anemic, and, after a single mating season, die. Females also have high mortality rates but often survive to produce more than one litter. Interestingly, the sex ratio among offspring in several *Antechinus* species is female-biased. Natural selection has worked out which sex, on average, has the greatest reproductive success.

How does the bizarre life history of the brown antechinus and its relatives match up to theoretical predictions? It seems likely that semelparity in males is driven by a high adult mortality rate among pregnant females. This mortality rate makes mating with only one female a risky undertaking and so favors multiple mating, even though the effort required to achieve it is fatal.[12] An interesting twist to the story is that there is some adaptive variation among populations. A study of a relative of the antechinus, called the dibbler, on two islands in Western Australia found that on one of the islands, where seabirds nested, the soil was eighteen times more fertile than on the other, which lacked seabird colonies. Dibblers eat insects, and their preferred food was more abundant on the more fertile island. On that island, male dibblers were in better condition after mating than those on the more impoverished island, and some of those males were not semelparous.[13] If the survival of pregnant females was also much better on the island with the richer soil because more food was available there, the theory would explain why semelparity in males would be less advantageous there too. Whether this is so is not yet known.

The capelin is a marine fish of subarctic waters that also has alternative life histories in two different environments.[14] Fish of both sexes that spawn in open water are semelparous, while those that spawn on the shore in the intertidal zone are not. This life history difference is maintained when both types are kept in a common aquarium environment, so it may have a genetic basis. The rule for semelparity would predict that the open-water breeders must be subject to high adult mortality, perhaps from predatory fish, while those that breed in the intertidal zone are relatively protected.

We tend to think of maternal care as something peculiar to birds and mammals, but it crops up among insects and spiders too, in which it often favors semelparity.[15] Females of the crab spider, for example, guard their egg clutch from predators for 40 days and 40 nights, during which time they lose 30 percent of their body weight, which is sufficient to prevent them from breeding a second time.[16] The female Japanese hump earwig lives under stones beside streams, where she guards her eggs until they hatch, whereupon the young eat their mother before dispersing from the nest. This behavior increases the survival chances of the young earwigs but only hastens the inevitable end for their mother, who would probably die soon anyway in the harsh environment she inhabits.[17]

High adult mortality due to environmental causes is one route to the evolution of semelparity, but another route is opened when concentrating reproduction into one big bang produces more offspring than repeat breeding would. This can happen through economies of scale. In the early days of the automobile, vehicles were produced in small numbers by craftspeople operating in much the same way that they had when they made horse-drawn carriages. Then along came Henry Ford. His huge factories were costly, and he paid his operatives well, but by using economies of scale he was able to churn out very large numbers of cars at an affordable price. The capital cost of such a production method is high, but the cost per unit produced is low. Many organisms make semelparity pay in a similar way. Pacific salmon are a classic example.

Pacific salmon belonging to several species spend half their lives in the ocean, where they lead chaste lives dedicated to feeding. Coho salmon, for example, after a year and a half of waking up every morning in the ocean with but a single thought—"Let us prey"—are fat and fit enough to breed. Fish then swim to the shore and enter a river, but not just any river.[18] Each fish seeks out the river in which it was born and heads upstream to the exact spot where shallow, well-oxygenated water and a gravel streambed provide just the right conditions for eggs to be laid and to survive. Salmon use Earth's magnetic field to guide them across the ocean, and in the final stretch they seem to use a memory of the taste of home waters to find their natal river.[19] But why do they head for home, rather than just swimming up the nearest river?

The salmon must make an arduous journey upriver, swimming against the current and evading predators. Some salmon were born near the coast and do not have far to travel, but others may have a truly epic journey of a thousand miles or more. The longer the river, the better prepared they must be before they attempt the journey. If they choose a river that is too long for them, they will die before reaching the breeding grounds. The only way that salmon can minimize this risk is to return to their natal river. Even within a river, breeding sites vary from small streams to broad, swiftly flowing waters, and these local conditions require a particular set of inherited adaptations. The migration, repeated generation after generation, with descendants inheriting the genes of the fish that choose the right spot on the right river and natural selection weeding out those that do not, ensures that evolution has honed each fish stock to survive its own journey home and to reproduce successfully there.

When the fish run, the rivers are so full of salmon that everyone, from Native American hunters to bears, takes their tithe of the bounty. The transfer of nutrients from salmon to the riverbank via the activities of predators is so substantial that the vegetation along salmon rivers in British Columbia is fertilized and permanently altered by it.[20] Predation has such a strong effect on the adult salmon mortality rate that this alone must favor the evolution of semelparity.[21] But in addition, for those few salmon that survive the journey home, the effort involved represents a huge investment that can be repaid only by a fatal effort to produce the largest possible number of eggs. Semelparous salmon produce a greater weight of eggs per unit of body weight than repeat breeders, and their eggs are also larger, which gives the young salmon fry a better chance of survival.[22]

Atlantic salmon have a migratory life history similar to that of their Pacific cousins, but the Atlantic species is a repeat breeder.[23] The reason for this difference is not clear. Both Pacific and Atlantic salmon expend a huge effort in migrating to breeding sites, and both meet stressful competition among breeding fish when they get there. Thus there is no obvious difference in the capital costs of breeding that could explain why Pacific salmon are semelparous and Atlantic salmon are repeat breeders. Part of the answer may be that the difference between their life histories is not as absolute as it at first appears. Although Atlantic salmon can breed more than once, and females in particular often do so, repeat breeding is rarer in males. In some rivers fewer than one in ten adult fish actually manage to return.[24] A factor that could account for the difference between Pacific and Atlantic salmon is mortality caused by predators during migration. No quantitative comparison has been made, but the feeding frenzy provoked by the annual run of Pacific salmon does not seem to have its equivalent on the Atlantic seaboard, which might mean that adult mortality is much greater in the former than in the latter.

Competition among female salmon for nest sites, and among males for access to spawning females, causes battles within the sexes, for which evolution equips them with pugnaciously hooked jaws. As the salmon return toward fresh water from their oceanic feeding grounds, the teeth they used in feeding are shed, and the lower jaw is transformed into a jutting weapon of war, terminating in a hook called a kype. The jaw and kype are particularly large in males, which use them in energetic battles for supremacy that are sometimes fatal. Fighting, though, is not the only way that male salmon manage to sire offspring. Even among Atlantic salmon, there are pacific males.

In fact, among Atlantic salmon and Pacific species such as the coho, there are two quite different types of males: the migratory, ocean-fed, hook-nosed bruisers; and much smaller, younger males that resemble juveniles, but that instead of migrating to the sea, remain resident in fresh water and become sexually mature there. These precocious gigolos in short pants are called jacks. They may skirmish among themselves, but jacks are not equipped to do damage. Instead, their mating strategy is to sneak fertilizations by hiding near a female's nest and then, when an opportunity arises, darting in to spill sperm there as she lays her eggs under the hooked nose of her chosen mate.

Each of these very different male strategies seems to be successful in its own way. Even though jacks avoid the heavy costs of ocean migration, they are unable to supplant hooknoses, because jacks depend on hooknoses to induce females to lay their eggs, so if hooknoses become rare, jacks lose out too. There is also an inherent limit on the reproductive success of hooknoses. As these males become more abundant, they fight more and more among themselves, which gives the jacks a relative advantage.[25] Because coho salmon are semelparous, jacks die at younger ages than hooknoses in this species. In repeat-breeding Atlantic salmon, jacks are delayed in migrating to the sea, and this increases the risk that they will never make it.[26] However you look at it, there is no escaping the cost that reproduction exacts on survival.

Two kinds of opportunities for economies of scale lead to semelparity in plants: escaping seed predation and attracting pollinating insects. Bamboos, which are wind pollinated, may delay flowering for a hundred years or more in some species before shunting all their stored resources into a humongous crop of seeds. Bamboos are grasses, so their seeds are as tasty and nutritious as wheat grains, and animals of all kinds gorge on this rare cornucopia of food when it occurs. If it were produced in small quantities at regular intervals, every grain would be quickly consumed, but by synchronizing crops and swamping predators with knee-deep supplies, the bamboos allow some of their seeds to escape. It is not known how bamboos achieve flowering synchrony, but there seems to be some kind of endogenous clock, because parts of the same clone that have been planted in different parts of the world have been known to flower and die in the same year.[27] Giant pandas feed exclusively on the leaves of semelparous bamboos, and occasional starvation in these threatened animals has been reported when they have been unable to move from an area where the bamboo has died to other areas that are out of sync.[28]

Periodical cicadas present the same kind of nutritious bounty for predators as flowering bamboos when they emerge in their millions from their protracted period of underground development to mate, lay eggs, and die. Some populations emerge after 13 years and others after 17, but these different broods are never found together. By emerging synchronously within an area, cicadas overwhelm their predators. So many emerge that when they die, their decaying bodies create a pulse of nitrogen in the soil that feeds forest plants.[29] In this case, experiments have revealed how the synchrony is achieved. Investigators tricked nymphs of the 17-year periodical cicada into emerging a year early by inducing an extra growth cycle in their host trees so as to create two periods of "spring" growth in a single year. Periodical cicadas count seventeen spring growth cycles in the trees on which they feed and then shout, "Shoot!" in chorus.

In some environments plants must compete with one another to attract pollinators to their flowers, and the biggest displays are the most attractive. In order to produce a very large floral display, a plant needs to store up resources and delay reproduction until it is capable of the big bang required to attain a suitably large economy of scale. The century plant *Agave americana*, which grows in Mexico and the deserts of the American Southwest, is one example among several semelparous species in the genus *Agave*.[30] Giant lobelias in the subalpine zone on Mount Kenya in Africa and huge herbs in the genus *Puya* in the high Andes also grow to massive size before, after many decades, shunting their all into a single giant flower display that natural selection has designed to impress the local insect and bird fauna.

Semelparity is an unusual but revealing life history. It is the ultimate demonstration of how costs of reproduction limit life span and of how conditions in the environment can lead to the evolution of extreme and paradoxical reproductive behavior. Most organisms are not semelparous, but repeat breeders are just as subject to the same evolutionary forces that kill or protect individuals and shape life histories, as we shall see in the next chapter.

8
Live Fast, Die Young
PACE

Every night I'm in a different town,
I'm the kind that likes to get around,
Living fast I'm on the run,
I take my chances cos' I'll only die young.

VENOM, "LIVE LIKE AN ANGEL (DIE, LIKE A DEVIL)"[1]

Live fast, die young is the defiant motto of a rock-and-roll life-style that is repeatedly inscribed in tattoo ink and printed in the obituaries of the prematurely dead. If rock musicians were a species unto themselves, and maybe they are, then biologists studying them would surely record the curious coincidence that many of the kind die at 27 years of age.[2] The gene or genius for dying at 27 seems to have originated with the granddaddy of the blues guitar, Robert Johnson (1911–1938). Pioneer of the electric guitar Jimi Hendrix (1942–1970) carried the baton for an equal stretch of years, and Janis Joplin (1943–1970), Queen of Rock and Roll, died a month after him, also aged 27. Both outlived Brian Jones (1942–1969) of the Rolling Stones, but only just. Jim Morrison (1943–1971) of the Doors died a year later, aged 27. More recently,

English R&B artist Amy Winehouse (1983–2011) died a couple of months before her twenty-eighth birthday.

Causes of death in the 27 Club are recorded as strychnine poisoning (Johnson), drowning (Jones), asphyxiation (Hendrix), heroin overdose (Joplin), heart failure (Morrison), and alcohol poisoning (Winehouse).[3] There are at least forty lesser-known members of this exclusive posthumous club who lived fast and died young. Rock musicians know in their guts that life span is governed by an unforgiving trade-off between the pace and length of life. Sad to say, some killjoy statisticians with nothing better to do have actually tested the hypothesis that rock musicians have a propensity to die at age 27, and they found the pattern to be illusory, at least for British pop stars.[4] The study did find, however, that mortality among musicians in their twenties and thirties was two to three times greater than in the population as a whole, so the idea that rock stars tend to die young is no myth.

But by comparison with some other mammals, rock musicians lead lives that are painfully protracted and slow. Ounce for ounce of body weight, a solitary shrew burns energy at twenty-five times the rate of a rock star. Or, to put it another way, a person's weight in shrews would produce enough energy to power a fantasy jam session by all twenty-five of the musicians that constituted the Rolling Stones, the Doors, the Jimi Hendrix Experience, and Amy Winehouse's band.[5] Shrews are small mammals that are possessed by a desperate need to consume sufficient food to fuel a furiously demanding lifestyle. They must eat two or three times their own weight in food every day, and 12 hours without food is enough to cause death by starvation. By comparison, a human can live for weeks on water alone. At the age of 74, the Indian social and political activist Mahatma Gandhi survived a fast that lasted 21 days.

Two things drive a shrew's almost insatiable need for food: small size and a dietary preference for insects. Mammals and birds are homeotherms, which means that their physiology is programmed to maintain a constant body temperature. Regulating body temperature, just like keeping your house warm in winter, involves balancing the internal generation of heat against its loss to the outside. Heat is generated by the cells of the body burning glucose and is lost by radiation through the surface of the skin. Body size affects this balance because the opposing processes of heat generation and heat loss scale differently as bodies get bigger.

If we imagine a body to be spherical (which requires only a small imaginative leap in the case of, say, dormice or certain opera singers), then the total volume of heat-generating cells is proportional to the cube of the body's radius, while the surface area through which that heat is radiated is proportional only to the radius squared. Now compare a tiny sphere that has a radius of, say, a tenth of an inch with a much bigger sphere with a 10-inch radius. The ratio of volume to surface area for the tiny sphere is 1:1, but the ratio for the larger sphere is 100:1. In other words, to achieve the same temperature, the tiny body must generate a hundred times the heat intensity of the larger one. What this amounts to is that a shrew has great difficulty keeping warm, while a whale has trouble keeping cool.

The only way that a small mammal can generate enough heat to stay alive is to continually stoke the boiler. This is true of all small mammals, but shrews have an additional handicap: their insect food is not especially rich in energy. Small rodents that feed on seeds have a much easier time of it than insectivores because seeds are packed with energy-rich compounds like fats and starches. Seed eaters are cooking with gas, insect eaters with a candle, but both are compelled by their small size to live fast.

Large animals, if they lived at the same rate as small ones, would burst into flames, and whales would boil the ocean around them with the heat generated by their metabolism. This does not happen because as body size goes up, the rate of metabolism falls. A shrew's heart hammers away at the incredible rate of more than 600 beats per minute, while that of an elephant is paced at a sedate 25 beats per minute.[6] In 1908, a German physiologist named Max Rubner (1854–1932) published the results of researches on the relationship between rate of metabolism and longevity that he believed revealed a golden rule: animals that live fast, die young.

Rubner measured the metabolic rates of five domesticated mammals ranging in size from a guinea pig to a horse and with life spans ranging from 6 years (guinea pigs) to 50 years (horses). Smaller animals have a higher metabolic rate than larger ones, but Rubner calculated that when you compare the total amounts of energy used over a lifetime, the short-lived guinea pig and the long-lived horse use about the same amount per ounce of tissue. This rule even seems to work when comparing shrews and rock stars. A shrew burns energy at twenty-five times the rate of a rock star, but typically lives for less than a year, so a rock star's cells take just over 25 years to burn through the same amount of fuel before joining the 27 Club.

Rubner's experiments on metabolism appeared to suggest that length of life might be determined in some way by a limit on energy consumption. If individuals of different species have roughly the same ration of energy that they can use during a lifetime, how long it lasts will depend on how fast it is used up. Rubner's idea is intuitively appealing if you think of the body as a machine that wears out more quickly the faster it is run, though we now appreciate why the machine analogy is a misleading one in the context of senescence (see chapter 6). But even if we accept the analogy, why should energy be rationed or the capacity of cells to use it be limited?

Rubner's rate of living hypothesis was given a big boost through the advocacy of Raymond Pearl (1879–1940), an influential American biologist and statistician who penned a prodigious output of some seventeen books as well as seven hundred articles that were published everywhere from the respectable *Ladies'*

Home Journal to the recondite *Proceedings of the National Academy of Sciences*. Everything he chose to write about, from cancer to cantaloupes and poultry to population growth, came from Pearl's deep conviction that whatever the problem, numbers were the solution. Unfortunately, he didn't always get the numbers or their interpretation right, and he could be extremely aggressive toward those who pointed this out.[7] His most disastrous error was a flawed analysis of the autopsy records at Johns Hopkins Hospital, where he worked, in which he drew the erroneous conclusion that tuberculosis prevented cancer. This analysis led to several terminally ill cancer patients being injected with a substance derived from the tuberculosis bacterium. Although the patients died, he regarded the treatment as a success.[8]

It is not surprising that length of life and rates of mortality should have interested Pearl's mathematical mind, for they lend themselves quite naturally to quantification. From age 24, when he published his first paper on mortality rates, to his sixteenth paper in a series called *Experimental Studies on the Duration of Life*, published a year after he died, Pearl pursued a mathematical solution to the laws of living and dying.[9] In 1919 he joined the medical faculty of Johns Hopkins University in Baltimore, but just three weeks after moving in, he suffered a huge setback when all his research materials, papers, and even the experimental mice he was preparing for a long-term study of aging were destroyed by a lab fire.[10] After the calamity Pearl picked himself up, issued an appeal to other scientists to help him restock his research library, switched to studying fruit flies, whose shorter lives yield quicker results, and drove right ahead with his work.

But Pearl's life wasn't all work. He was an enthusiastic member of the Saturday Night Club, whose members caroused at the Baltimore home of the satirical writer and journalist H. L. Mencken. The club's heraldic shield sported sausages, a lobster, a beer stein, and a fiddle, all garlanded with onions and pretzels. The music at the club was provided by students from the local conservatory, with Pearl joining in on French horn. It wasn't rock and roll, but it was full-on. On one occasion the club's orchestra planned to play the first eight of Beethoven's symphonies in a row. Pearl had reached the first movement of the fifth symphony when his French horn reportedly "blew up."[11] It would have been typical of Pearl to have then conceived a study of the mortality rate of wind instruments, or at least their players, but this seems to have been one of the few statistical projects that never occurred to him.

These were the days of Prohibition, so the Saturday Night Club's beer was brewed clandestinely in Mencken's cellar, where bottles regularly exploded under the pressure of fermentation.[12] Pearl appears to have been the first scientist to investigate the effects of alcohol on mortality, even studying its effects on the growth of seedlings.[13] He discovered that moderate alcohol consumption did not shorten life, a finding that has been confirmed by more recent studies, which suggest that modest imbibing can lengthen life.[14] Later Pearl was among the first to demonstrate that even moderate smoking was harmful to longevity,[15] leading him to the wry observation that he was going to stop smoking and drink more.

Pearl dedicated his 1926 book *Alcohol and Longevity* to the members of the Saturday Night Club,[16] who must have been glad to read its conclusions through the bottom of their beer glasses. Such a dedication may seem like a daring challenge to the authorities in the midst of Prohibition, but Pearl was well known at this time for being independent-minded because his more journalistic writings often used science to debunk myths. Pearl even had a walk-on role in a Pulitzer Prize–winning novel published in 1925 by Sinclair Lewis: Dr. Martin Arrowsmith, the fictional and eponymous hero of the novel, consults Raymond Pearl, who is characteristically skeptical about the validity of Dr. Arrowsmith's evidence that he has found a cure for bubonic plague.[17]

Pearl's experimental researches on the relationship between rate of living and longevity were done with fruit flies and cantaloupe seedlings. Like others before him, he found that flies kept at low temperatures lived longer than those kept warm. Chilly flies do not move much, so Pearl concluded that lack of activity extended life. His cantaloupe seedlings were grown in the dark and starved of sustenance. Pearl was nothing if not ambitious for the profound significance of his simple experiments. So when he found that seedlings that grew more slowly lived longer, he interpreted this finding as more proof of the general principle—live fast, die young.

In his book *The Rate of Living* (1928), Pearl concluded that all the evidence pointed to the fact that "the length of life depends inversely on the rate of living."[18] Pearl believed that this universal rule of life explained variation in length of life among humans too. However, he cautioned the scientific audience that read or heard his lecture series, "The Biology of Death," that data on the relationship between people's occupations, their energy expenditures, and how long they lived were almost impossible to interpret and could not be used to prove the theory.[19] Yet only a few years later, he was happy to headline a popular article in the *Baltimore Sun*: "Why Lazy People Live the Longest."[20] Pearl himself lived only to the age of 61, though it's an interesting thought that he might have lived longer had he been too lazy to write such tosh.

Even if Pearl was a tad overenthusiastic in promoting the rate of living hypothesis, evidence in its support began to accumulate from other quarters. The beating heart of a water flea is visible through its almost transparent body, so researchers were able to count the heartbeats of these tiny crustaceans as they lived out their lives in containers kept at different temperatures. Sure enough, the cooler water fleas lived longer than the warmer ones in exact proportion to the slowing of their heartbeats[21]—more confirmation for the rate of living hypothesis. So exact was the inverse relationship between heart rate and life span that the little creatures might almost have cribbed the answer from Pearl's book. Data on metabolic rates were collected for more species of mammals, filling in the gaps between guinea pig and horse and stretching to include smaller and bigger animals, proving that the relationship that Max Rubner had discovered really was a general one.

By the 1950s the rate of living hypothesis looked pretty solid, and the big question was: What is it that sets the limit on lifetime energy consumption and thereby limits length of life? Pearl had believed that cells must contain some vital molecule that gets used up, though what it was, was beyond even his ability to calculate. Then, in 1954, a physician working at the University of California, Berkeley, by the name of Denham Harman came up with a different idea. He was puzzled by the universality of aging, and having worked for 15 years as a chemist for Shell Oil before going to medical school, he was well prepared to think about the problem from a chemical angle. After four months of worrying continuously over the problem, the answer finally just popped into his head.[22]

Harman proposed that limits on life span were set not by the exhaustion of a compound, as Pearl had speculated, but by the accumulated damage caused by a particular type of molecule produced in metabolism. The culprits were molecules called free radicals, which are produced whenever chemical energy is released from sugars through combination with oxygen. Chemically, this reaction, called aerobic respiration, is a highly controlled process of burning. Perhaps the idea that aerobic respiration, like any combustion in air, would produce dangerous by-products was likely to occur only to a chemist like Harman who had worked for an oil company. This might be why, for nearly a decade after he published it in 1956,[23] the idea was either ignored or ridiculed by biologists who did not seem to understand the chemistry. It was more than 20 years before the idea began to catch on, and then it really caught fire.

Free radicals are small molecules that have an unpaired electron. Electrons are negatively charged particles that prefer company, which makes free radicals chemically very reactive. The type of free radical that Denham Harman thought was causing trouble in cells contains an oxygen atom with an unpaired electron. Oxygen free radicals are potentially damaging because they attach themselves to molecules in the cell, oxidizing them and stopping them from performing important biological functions. Household bleach is an oxidizing agent that has some familiar effects on biological substances, or stubborn stains, as the TV ads prefer to call them. Imagine free radicals with that kind of oxidizing power inside a cell and you get some idea of the damage they might do. Oxygen free radicals can damage practically any of the important molecules in a cell, including fats, proteins, and the nucleic acids of which DNA and RNA are made. Damage to DNA does accumulate with age, but whether this damage is the most important cause of aging, as some scientists have argued, is not clear.[24]

Harman's free radical theory of aging supplied the missing piece in the rate of living hypothesis. The two theories meshed together like well-oiled cogs in an efficiently operating machine. The rate of living hypothesis proposed that life span is inherently limited by the deleterious effects of metabolism. Run the living machine at 600 heartbeats a minute and death will soon follow. Slow the pace of life to that of a large mammal, and the grim reaper will delay his visit. The free radical theory explained how the rate of living could have such effects on life span. Aerobic respiration is a pact with the devil: you certainly can't live without it, but you can't live with it forever, either. Every calorie you burn in the fire of life also stokes your funeral pyre. Curiously, this is not a new idea. William Shakespeare wrote a sonnet in which he compared old age to the glowing embers of a fire:

That on the ashes of his youth doth lie,
As the death-bed whereon it must expire
Consum'd with that which it was nourish'd by.[25]

By the end of the twentieth century, the two theories of aging had effectively become one. Biologists had explored every nook and cranny of the well-oiled machine, uncovering its living mechanism in molecular detail.[26] An important turning point for Harman's free radical hypothesis was the discovery in 1969 of an enzyme in cells that converts one of the most potent oxygen free radicals into less harmful molecules. The enzyme was named SOD. A host of other antioxidants were also found, including more enzymes and some small dietary antioxidant molecules derived from fruit and vegetables. This armory of cellular defenses against free radicals amounted to a strong endorsement from the highest possible authority—nature herself—of Harman's idea that they were dangerous. Perhaps this endorsement ought to have rung an alarm bell, however. If cells are naturally so well defended against free radicals, it shows that Harman was right about the potential threat they pose, but he might not have been right about their actual contribution to aging, because maybe nature has got the problem covered.

Meanwhile, back at the farm, biologists were taking a closer look at the relationship between metabolic rate and longevity in some animals that didn't seem to have read Raymond Pearl's instructions. In 1991 Steven Austad and Kathleen Fischer, then at Harvard, noticed that most of the species that fit the known relationship were terrestrial, placental mammals. Did other kinds of mammals also fit? Bats live about three times longer than terrestrial mammals of the same body size, while marsupial mammals like kangaroos and opossums have lives that are about 20 percent shorter. According to the rate of living hypothesis, then, bats should have a slower metabolism than shorter-lived terrestrial mammals, and marsupials should have a faster metabolism, consistent with their shorter lives. What Austad and Fischer actually found was quite different. Bats had metabolic rates similar to those of terrestrial mammals with similar sizes and shorter life spans, while marsupials had slower metabolic rates than expected for their body size and length of life. Not only that, but among species of bats and marsupials that conserve energy by hibernation, there was no evidence that their lives were made any longer as a consequence.[27] Birds show an even more deviant pattern than bats, being longer lived than and having metabolic rates double or more those of terrestrial mammals.[28]

The final nail in the coffin of the rate of living hypothesis was delivered by João Pedro de Magalhães at Harvard, who compiled a database called AnAge on longevity in animals, which at the time of this writing covers more than four thousand species.[29] In 2007, almost exactly a century after Max Rubner's pioneering study started the hare running, a comprehensive analysis of the data in AnAge showed that there is no correlation between longevity and metabolic rate among birds or placental mammals if animal size is taken out of the equation.[30] In other words, Pearl's rate of living hypothesis was built on a false premise. The relationship he thought existed between longevity and metabolic rate was in fact due to a relationship between metabolic rate and body size; it was also biased by the narrow range of species he was able to study.

But what of the experimental evidence that Pearl marshaled in support of the rate of living hypothesis? This fares no better. Simple experiments like Pearl's can have genuinely profound implications, but Pearl's conclusions were simpleminded because he ignored alternative possible interpretations of his results. In the experiments in which he cooled fruit flies, he forgot that all biological processes tend to be slowed by cooling. Therefore, showing that two processes—activity and mortality—are both slowed in chilly flies does not prove that a slowdown in one causes the observed slowdown in the other. Some unknown aging process that is also slowed by cooling, but that is unrelated to activity, could be the real cause of longer life in the fly experiment. This is known as the "third variable problem," and as we have seen, Pearl and many others were caught out by it in their analyses of the relationship between metabolic rate and life span: in that case the third variable was the very obvious one of body size.

In the case of Pearl's cantaloupe experiment, it was conducted in such highly constrained, artificial circumstances that it revealed nothing more profound than that under starvation conditions, the rate at which finite resources are used up by growth determines how long seedlings will last. Recall that in the more recent plant experiments described in chapter 5, fast growth also raised the mortality rate, but this occurred only under conditions of stress. "Live fast and die young if times are bad" is not the universal law that Pearl was looking for, but it does hint at something important that we have neglected up to now: the condition of the environment and its effect on life span.

The rate of living hypothesis is dead. Metabolic rate does not determine length of life. However, body size, with which metabolic rate is correlated, does seem to influence life span, so let's look more deeply into that. As we saw in chapter 2, the idea that body size influences life span goes back to the ancient Greek philosopher Aristotle, but there are some clear exceptions to the rule that bigger animals live longer. Those exceptions turn out to be quite informative. The naked mole-rat, for example, is a tiny rodent that lives two to three times longer than the biggest member of rodentkind, the capybara. Mole-rats, as their name suggests, live underground, where they are well protected from many predators (but not some snakes), and this lifestyle does seem to be associated with longer life in subterranean mammals generally.[31]

As already noted, both bats and birds typically live longer than nonflying animals of the same body size. Could flight and the protection from predators it provides be responsible for this difference? A good test of this hypothesis is afforded by flightless birds, which during their evolution have traded the ability to fly for large body sizes. This evolutionary transition occurred independently several times on oceanic islands that were predator-free until humans and rats arrived. An example is the dodo, that icon of extinction, which was a turkey-sized, flightless pigeon that evolved on the Indian Ocean island of Mauritius. Has the loss of flight shortened the lives of such birds? We do not know what the life span of the dodo was, but there are two other large flightless birds still living in which life span has been measured. The ostrich can grow to over 200 pounds and lives for up to 50 years in captivity. This is a respectable age for an animal of its size, but not at all remarkable for a bird. African grey parrots weighing only a pound live at least as long. The other flightless bird we can compare is the emu, which can grow to 80 pounds, but which lives for only about 17 years. Its life span is the same as that of the American robin, but that species weighs only 2.5 ounces. Clearly, size is not everything, and the power of flight does seem to extend life.

In addition to large body size and the ability to hide underground or fly away, other features that are associated with longer life include chemical defenses that make an animal unpalatable,[32] hibernation,[33] living in trees if you are a mammal,[34] and the body armor of turtles.[35] What are these disparate features telling us? The single explanation that seems to account for the positive effect of body size and such a miscellany of other characteristics on life span is that they all offer protection from predators. Remarkably, George C. Williams predicted exactly such a pattern in his paper on the evolution of senescence published in 1957, at a time when nearly everyone else was taken with the flawed rate of living hypothesis.[36]

Williams's argument went like this: Natural selection will favor the kind of organism that leaves the most offspring, so to work out how many offspring any particular kind of organism will leave, divide the life span into compartments representing successive ages and add up the number of offspring produced in each compartment. It's a simple enough idea, but I'm going to introduce a metaphor that will help explain some of its important intricacies.

Life span is represented by a line of railroad carriages, and each carriage is a period of life. (I could say that each is a carri-age, but I won't!) At the front of the train is the engine, which represents the juvenile period of life. The engine carries no passengers itself, but its running is critical to the fate of the passengers in all the carriages that follow. Think of the passengers as offspring, and you can see that they will go nowhere if the engine dies. We are interested in how long the train, or life span, is going to be, so let's take the engine for granted. Natural selection, when it operates on life span, is just as uninterested as we are in any engines that fail to leave the station, because they carry no offspring.

Behind our functioning engine is a line of ten carriages, coupled to one another by rather fragile chains that are liable to break at random. The first carriage represents the youngest adult age class, the tenth represents the oldest. The operator of this train is natural selection, and it loads passengers/offspring into all ten carriages before the train sets off. Halfway to the next station one of the couplings breaks, and all the offspring (let's drop the passenger pretense) in the carriages to the rear of the breakage are lost. If you were in charge of loading such a train and wanted as many offspring as possible to reach the next station, to which carriages would you allocate them?

This question is easy to answer: use the younger carriages and avoid the older ones. The tenth carriage, representing the oldest age class, has the greatest chance of being lost because there are nine other carriages and ten fragile couplings between it and the engine. If every coupling has the same likelihood of failing, offspring in the last carriage have ten times the risk that they will not make it compared with offspring in the first (youngest) carriage. This argument is the same one by which Peter Medawar deduced that older ages matter little to natural selection because they make so small a contribution to the next generation.

You might have answered the question I posed by saying, "I'd put all my offspring in the first carriage." That would certainly be the safest thing to do if it were possible, but the carriages are of limited size, and you have a lot of offspring to shift, so you are going to have to spread them out. Now the question is, how far down the train will you risk placing your offspring? The answer to this question will depend on how great the risk is that couplings will fail. If that risk is high, then you might treat the train as if it were very short and load only the first few carriages. A short train is like a short life span. If the risk is lower, more carriages can be safely loaded, and the train, and the life span it represents, will be longer.

All that is now needed to complete this metaphor is to reveal that the unreliable couplings represent the extrinsic mortality risk to adults. "Extrinsic" simply means that the source of the risk is outside the control of the organism itself, although it can hide from it, flee from it, or defend itself through sheer body weight. Returning now from the metaphor to reality, we see why Williams predicted that high adult mortality favors a short life span while lower mortality favors longer life.

One cautionary whistle needs to be sounded. It's a warning to look up and down the line before you venture to cross the railroad tracks, lest you be mown down by my runaway metaphor as she goes over the hill. I've described the evolutionary process that determines length of life as though natural selection chooses in some purposeful, goal-directed way how many carriages to fill, but that idea is misleading and is not to be taken literally. Natural selection is blind to how the carriages are filled and does not count offspring till they arrive safely at their destination. Rather, there are a thousand, a million, or a billion trains, and those that deliver the most offspring are the kind that are replicated at the next station, while the trains longer or shorter than the optimum grind to an ignominious halt in the sidings of oblivion.

Compare the rate of living hypothesis with the mortality hypothesis just described. An important difference between them is that the latter deals quite explicitly with how life span is shaped by natural selection, while the former is silent on that issue. The mortality hypothesis predicts that there is a continuum of life-history types, from short-life-span fast trains that speed quickly from one generation to the next to long-life-span slow trains that take their time between the generation stations.

The rate of living hypothesis predicts that life span is inversely related to metabolic rate, but this prediction has been refuted by the discovery that animals such as birds have long lives and a fast metabolism. The mortality hypothesis says that "live fast, die young" is indeed a rule of life, but that what runs "fast" or "slow" is not metabolism, but the speed of the life cycle as measured by generation time. By this definition, bats and birds have slower life histories on average than terrestrial mammals, not faster ones.

There is a problem with testing the mortality hypothesis that may already have occurred to you: Isn't it inevitable that populations exposed to greater mortality will live shorter lives? Yes it is, so what a proper test needs to show if the hypothesis is correct is not just that there is a correlation between adult mortality rate and length of life, but that evolution has altered senescence in response to the impact of mortality.

Measuring rates of senescence in wild populations requires that detailed information be collected on many individuals over long periods, so these data are harder to obtain than simple statistics of longevity. However, the available data for birds and mammals do show, as predicted, that senescence is more rapid in populations exposed to high mortality,[37] and that birds and terrestrial mammals with the same generation times senesce at the same rate.[38] These findings suggest that if terrestrial mammals have faster life histories on average than birds, it must be because generation time is shorter on average in the mammals, and that, in turn, must be because adult mortality is higher on average in the mammals.

So far, so good for the mortality hypothesis, but correlations, as we discovered in the case of the now discredited rate of living hypothesis, can be as misleading as they are beguiling. It would be nice to have some experimental evidence, and, as ever, fruit flies are happy to oblige. Fruit fly experiments are conducted in half-pint glass bottles of the kind that milk was once sold in. A standard amount of food is placed in each bottle, and a fixed number of fly eggs is counted into each. The eggs hatch to produce tiny larvae, which tunnel through the food medium as they devour it. A week later the larvae are full grown, and they climb the sides of the bottle, where they fix themselves to the glass and pupate. Inside the pupal case an extraordinary transformation takes place as the tissues of the larva turn to goo and rearrange themselves into the complex structure of an adult fly. In another week this amazing metamorphosis is complete, and the adult emerges.

You might think that this sounds like such a simple experimental system that nothing much could go wrong, but the problem is to design experiments that give an unequivocal answer, even to a question as seemingly straightforward as "Will an increase in extrinsic adult mortality cause a shortening of life span?" The difficulty is ruling out those pesky hidden third variables. So, for example, the experimenter could cull flies from a set of bottles to impose increased mortality, but that treatment would lower the density of fly populations in the bottles receiving it, introducing an unintended decrease in crowding as well as the intended effect of higher mortality. Any effect or, indeed, lack of effect on fly life span could then be due to the change in population density, the change in mortality rate, or a combination of the two.

Problems of this kind plagued the interpretation of some of the earliest experiments, so it was not until the end of the twentieth century that all the experimental wrinkles had been ironed out and the mortality hypothesis was really put to an unequivocal test.[39] Experimenters culled flies from bottles twice a week, and after each cull, they maintained population density by topping up with fresh flies. In a low-mortality treatment flies had a 64 percent chance of surviving for a week, while in a high-mortality treatment the chance was only 1 percent. The experiment was run for more than fifty generations, which in human terms would be equivalent to about 1,000 years. After this many generations, would flies in the high-mortality treatment have evolved shorter life spans? This question was addressed by removing flies from the experiment, letting them lay eggs in new bottles, and measuring the natural mortality of the emerging flies over 100 days. Here, too, population density was kept constant by replacing dead flies with fresh ones. When flies were counted, the "extras" were separated from the flies being tested by a difference in eye color.

As predicted by the mortality hypothesis, life span was significantly decreased by the application of high extrinsic mortality, although the change brought about by fifty generations of selection was quite small. Flies from the high-mortality treatment had life spans that were on average only about 4.5 days, or 7 percent, shorter than those of flies that had been kept in conditions of much lower extrinsic mortality risk. The flies in the high-mortality treatment also changed their pattern of egg laying, reaching a peak at an earlier age than flies in the low-mortality treatment.

It seems very unlikely that the shorter life span of flies exposed to the high-mortality regime was due to the accumulation of new, late-acting mutations in just fifty generations. Instead, natural selection must have favored genetic variants with earlier reproduction. As we have seen in chapter 6, reproduction often carries a cost in terms of later survival, so earlier reproduction in flies exposed to the high-mortality treatment may itself have caused the shortening of their lives.[40] This is certainly how the life span of weeds adapts to conditions of high mortality, since the absence of a separate soma in plants rules out Medawar's mutation theory of senescence.

The definition of a weed is simply "a plant in the wrong place," which for most purposes is an arbitrary label, because who is to say what the "wrong place" is? However, the label "weed" is a very good indicator of the extrinsic mortality rate imposed on these plants by gardeners. Studies comparing the life spans of two very common weeds, groundsel and chickweed, discovered that the life spans of these species were significantly shorter in several assiduously weeded botanic gardens in Britain than in populations of the same species growing in natural habitats. Gardeners seem to have performed an unwitting experiment on the weeds, and by applying higher extrinsic mortality, they have selected for earlier flowering and shorter life span.

Adaptation to different mortality regimes also seems to account for differences in flowering time between two varieties of gentian that until quite recently were thought to be different species. One of them, the early gentian, is spring flowering and lives in closely grazed habitats, where it can complete its life cycle in only 14 weeks. The other, autumn gentian, lives in taller, less disturbed grassland, and it flowers in the fall of its second year. Studies of the genetics of these two plants have found that they are so similar that they really ought to be classified as a single species.[41] It seems that the short-lived early gentian is merely a form of autumn gentian that has evolved a shorter life span due to the high extrinsic mortality caused by grazing in the habitats where it is found. This discovery has important implications for plant conservation, since it was previously thought that the early gentian was one of the very few plant species that could be described as unique (endemic) to the British Isles. Perhaps British botanists will find consolation in the knowledge that they have the fastest gentians in the west.

Other examples of life span evolving to suit local conditions abound among animals as well as plants. Steven Austad, who showed that the rate of living hypothesis was contradicted by the metabolic rates of marsupials and bats, discovered an example that can be thought of as a natural experiment on the effect of adult mortality on life span. While working in South America, he noticed that the opossums he was studying there seemed to age at a shocking rate. "I'd catch them; they'd look great. They were healthy adults. I'd catch them three months later and they looked horrible. They had parasites. They had arthritis. They had cataracts. They were falling apart."[42] Could high predation rates on opossums account for their rapid aging? Austad had the idea that if he could find a population of opossums that had been protected from predation for many generations, he could test the prediction that animals living under a lower risk of extrinsic mortality would age more slowly. So he searched for an island where opossums were present but large predators were absent. Eventually he found what he was looking for in Sapelo Island, lying about five miles off the Georgia coast.

Previous surveys of the fauna of Sapelo Island had established that large predators such as pumas, foxes, and bobcats did not occur there. The first thing that Austad noticed was that the island opossums did not show normal predator avoidance behavior. Mainland opossums are nocturnal, but these animals wandered around during the day, and they would snooze on the ground without troubling to hide in an underground burrow as mainland opossums invariably do.[43] The animals were easy to catch, mark, and release, and as the data began to accumulate, Austad was excited to discover that their rate of aging was about half that of the mainland animals that he monitored for comparison.[44] Mainland opossums bred once, producing a large litter; only rarely did they breed a second time, and then with low success. Sapelo opossums had smaller litters, but they often bred a second time, and with no loss of fertility. This difference is exactly what is predicted by the extrinsic mortality hypothesis.

Live fast, die young—and its corollary, live slow, die old—appears to be a rule that all organisms have to live by. The speed of life has little or nothing to do with the rate of metabolism and everything to do with the pace of the generations. That pace is tuned by the hazards of adult life. The human species takes life very slowly, even by the idle standards of our fellow primates. Why should evolution have cut us such slack? The mortality hypothesis would predict that the answer must lie in the ability of our early ancestors to escape the high adult mortality rates that are typical of mammals as a group. Primates are tree dwellers, and this lifestyle is associated with longer life in all the mammals that share the habit.[45] We therefore started with an advantage that we later carried with us when our ancestors gave up tree living. Another pattern that is found right across the mammals is that species with bigger brains live longer.[46] Hence our slow lives must also owe something to our quick wits. Those quick wits are also responsible for the biggest leap in longevity our species has ever made: the doubling of human life span in the last 200 years. Do we now have the wit and the science to outdo Tithonus and to be forever young?

9
Forever Young?
MECHANISMS

May your hands always be busy
May your feet always be swift
May you have a strong foundation
When the winds of changes shift
May your heart always be joyful
And may your song always be sung
May you stay forever young

BOB DYLAN, "FOREVER YOUNG"

Bob Dylan wrote these famous lyrics for one of his children. Viewed from the prospect of age, the supple, unblemished beauty of youth is a wondrous thing, perhaps most especially to a parent. Nothing reminds us so keenly of the power of biological processes to reset the lapse of time as the pristineness of a child, freshly minted from the germ line. How cruel that as we age, our somatic selves must suffer the accumulated payback for youthful fecundity.

For centuries philosophers have dreamed of finding an elixir of youth so that they might remain forever young, but none remotely understood what aging was, or why it occurs, so they had not the slightest hope of conquering it. Now that we understand not only how biological function deteriorates, but also why, does this scientific knowledge offer new hope, or does it merely tragically rekindle a long-held delusion?

In Robert Heinlein's science fiction novel *Methuselah's Children*, a nineteenth-century millionaire by the name of Ira Howard who finds that he is prematurely aging uses his fortune to create a foundation whose task is to discover how to prolong human life.[1] After Howard's death the foundation adopts a breeding program in which it identifies the offspring of long-lived families, encourages them to marry each other, and rewards them financially for each child of their union. The incentive scheme continues generation by generation until, by the time the story opens, members of the Howard families have natural life spans of more than 200 years, although they appear quite normal. The families have to engage in subterfuge to hide their real ages from the general population of "ephemerals" who have normal life spans, but this becomes more and more difficult to carry off. When some of them reveal their real ages, the ephemerals refuse to believe that such extraordinary longevities result from generations of selective breeding and accuse the Howard families of selfishly hiding a secret elixir. The ephemerals want a quick fix for aging and cannot believe that such a thing does not exist.

This situation is analogous to where we are now in the science of longevity. You can have genes that favor long life, if you have the good fortune to inherit them, and there are worms, flies, and mice that can be genetically manipulated to have longer lives. It is abundantly clear that evolution has extended life span in some species and abridged it in others. We humans have been beneficiaries of this process of natural selection, as we live longer than any other primate. Through economic, social, and medical advances, average life span has advanced by nearly 15 minutes per hour over the last two centuries.[2] Yet we are not satisfied with this progress and demand an elixir.

Health food stores are piled deep with dietary supplements containing antioxidants and other substances that, it is claimed or implied, slow aging. Denham Harman himself suggested in his original 1956 paper that the damage done by oxygen free radicals might be reduced by feeding cells with an antioxidant molecule that would mop them up. It was an idea that was way ahead of its time, but not one that 60 years later justifies the billions of dollars that are spent on antioxidant dietary supplements. Numerous clinical trials of the efficacy of antioxidant dietary supplements such as vitamins A, C, E, and beta-carotene have failed to show any clear benefits, and in some cases the trials have even uncovered health risks.[3]

Since antioxidants are naturally present in a balanced diet anyway, one conclusion from these trials might be that nature has already got the problem of free radicals adequately covered. Furthermore, we now know that oxygen free radicals are not just dangerous by-products of metabolism, but actually serve a number of vitally important functions, for example, in growth and development and in the immune system.[4] Denham Harman's basic idea that oxygen free radicals are potentially damaging was correct, but it is now clear that this is far from the whole story and that the amount of damage, or oxidative stress, that they cause is regulated in the body. As ever, when you get down to mechanisms in biology, it's complicated.

One reason that it's complicated is that there are a variety of ways in which oxidative stress can be dealt with, and it seems that different organisms deal with it differently. For example, one study found that in the ocean quahog, the clam that claims the longevity record for animals, some tissues produce lower amounts of oxygen free radicals than are found in the much shorter-lived hard clam, but others do not.[5] While the same study found that the ocean quahog was more resistant to oxidative stress than the hard clam, it also found that there was no difference between the species in the activity of antioxidant enzymes such as SOD.

The longer-lived species was more resistant to oxidative stress than the shorter-lived one, but exactly why was not clear.

The tiny cave-dwelling olm salamander, or human fish, which is only finger-sized but lives for a century, does so without an unusually high level of antioxidants.[6] Nor do naked mole-rats, the rodent Methuselahs, have any special protection against oxidative stress, but they live ten times as long as the healthiest mouse, despite accumulating high levels of damage to their DNA and proteins caused by oxidation. They seem to tolerate these levels of stress by preventing damaged cells from proliferating.[7] Most damaging of all to the oxidative stress hypothesis are lab experiments that have shown that genetic manipulations of levels of antioxidants in mice and nematode worms affect levels of oxidative stress, but have no effect on how long the animals live.[8]

At first sight, this evidence would seem to kill stone dead the idea that oxidative stress might influence longevity, but lab experiments have a limitation that we have met before and should not forget. The first genetic experiments that uncovered genes that could prolong the lives of nematode worms seemed to show that they did so without any negative consequences or trade-offs with reproduction (see chapter 2). Later it transpired that when worms were raised under more natural conditions than a petri dish, the long-lived mutants were rapidly replaced by the shorter-lived wild type. So, by analogy, it is one thing to show that the presence of a high level of oxidative stress does not impair survival in the lab, but quite another to show that this is also true in the wild.

Birds are a particularly interesting model in which to test for the influence of antioxidants on survival in the wild because a class of antioxidants called carotenoids provide the red, orange, and yellow pigments that color some species' plumage, hence these animals need significant quantities. Carotenoids also give the yolk of an egg its color. Despite the importance of carotenoids, all animals lack the biochemical pathways to make their own, so these molecules, like the vitamins, have to be obtained from the diet. In birds, as in other animals, it is the male that tends to be the more brightly colored of the sexes (think of the showy peacock versus the dowdy peahen, for example) and the female that is choosiest in mating. Could it be that in species that use carotenoids to color their plumage, the brightness of their feathers signals which males are best fortified with antioxidants, conveying information that females can use to select the best father for their offspring?[9]

Carotenoids are only weak antioxidants, so while it is an attractive idea that male birds might be using the pigment properties of these molecules to signal how desirable they are as mates, it's a hypothesis that could quite easily fall apart. This is what makes the results of a test with the common yellowthroat all the more fascinating. Common yellowthroats are small perching birds found throughout most of the United States in summer. Males have a particularly bright yellow patch on the throat. In a study population near Albany, New York, this patch was found to be brightest in the healthiest males, which were also the males that females preferred.[10] Crucially for the hypothesis under test, males with brighter throat patches had lower levels of oxidative damage to their DNA, and those birds with less DNA damage survived the winter better.

Antioxidant levels have also been found to be correlated with survival in other wild bird populations. Each year barn swallows undertake a grueling migration from their nest sites in Europe across the Sahara to southern Africa, where they overwinter before returning faithfully to the same nest sites in the north. A 5-year study of three Italian colonies found that swallows of both sexes that had higher antioxidant levels in their blood survived significantly longer than birds with lower levels.[11] A study of the closely related American barn swallow has found that reproductive success is correlated with carotenoid concentration during the breeding season.[12]

It is wisely said that one swallow does not make a summer, but these field studies might spell a reprieve, or maybe just a temporary suspension, of the death sentence for the oxygen free radical hypothesis. They should also make us pause to consider whether the causes of aging are likely to boil down to a single mechanism or to be many and varied among species.

As with all fundamental questions having to do with senescence, it is worth knowing what G. C. Williams thought of this matter back in 1957, because so many of his predictions have turned out to be right. He argued that "senescence should always be a generalized deterioration, and never due largely to changes in a single system."[13] His reasoning can be understood if we bring the metaphorical life-span train back into service. Recall that each of the passenger carriages represents a successive age class during the life span. The carriage nearest the engine is the youngest, and each of those farther back is incrementally older. Previously, we assumed that all the couplings that join one carriage to the next have the same chance of breaking, and that this chance represents the risk of mortality due to extrinsic causes. Now let's relax that assumption and acknowledge that the construction of the couplings themselves will have some effect on the chance that they will come apart. In fact, let's say that each coupling is made of a chain that has four links in it.

Metaphorically, each of the four links represents a different biological system, every one of which is vital to survival beyond a certain age. For example, one link might represent the immune system, another resistance to cancer, a third resistance to oxidative stress, and a fourth efficient insulin signaling. A chain is only as strong as its weakest link, so all the links must hold if the carriage is to survive the journey. Now let's say that senescence is represented by the thinning of the metal from which a link is forged. Among the younger carriages each link is stout and strong, but those connecting the older carriages to the train are made of thinner and thinner steel, because as we already know, the offspring in the older carriages make very little contribution to future generations, so natural selection cares little or nothing about maintaining them.

Now let's get down on the track with the maintenance crew and inspect what's happening to the couplings around the middle of the train. These carriages represent middle age, and natural selection is beginning to lose interest in them, but she can still squeeze some use out of them. Oh, look! One of the four links in the coupling is much weaker than the other three. We check a few other trains, and there is no doubt about it. It's always the same link of the four—resistance to oxidative stress is beginning to weaken.

If the maintenance crew is being directed by natural selection, what should it do? Obviously, the best strategy is to fix the problem by strengthening the weakest link. This was the gist of Williams's argument: if any one vital system begins to weaken before the others, natural selection will strengthen that system. The existence of defenses against oxidative stress proves this point. Natural selection invented SOD and other mechanisms to fix the problem, not perfectly, but to the point where it is not the sole and universal cause of senescence. Any vital link that begins to give out consistently before the others will be subject to the long-term attention of natural selection. Then, at the point in the life span where natural selection loses its power altogether, anything goes, and everything does. That is why the biggest risk factor for nearly every ailment in the medical encyclopedia is the age of the patient. It's senescence.

This argument has important implications for the idea that we might find an elixir of life that will "cure" aging. If aging were one thing, a cure might be possible, but it isn't. It's a general failure of multiple systems. As such, the best that our evolutionary heritage will permit us to do is to lengthen the train and delay senescence; we cannot abolish it altogether. In the end everything except the germ line senesces and dies, including all known longevity mutants.[14]

There are scientists who think that aging can be conquered piecemeal, by fixing one system at a time, and they point to the apparently negligible senescence of animals like the ocean quahog or the olm as evidence that this should be possible. I'm tempted to say that, personally, I'd rather live a short life as a human than a long life as a human fish, but that's a cheap shot. Or is it? Trade-offs are universal, and it's unlikely that negligible senescence would have no downside.

The most proselytizing of the optimists who believe that aging can be cured is Aubrey de Grey, a maverick from Cambridge, England, the subject of a biographical account by science writer Jonathan Weiner.[15] De Grey describes his approach as employing "Strategies for Engineered Negligible Senescence" (SENS), and his aim is to find ways to repair the damage that accumulates as the efficiency of the normal cellular repair processes decreases with age.[16] De Grey believes that there are seven types of damage that require repair. Two of these are caused by mutations that damage DNA, including those that trigger cancer; two involve cells malfunctioning in various ways; two are due to the accumulation of toxic aggregates such as the plaques seen in the brains of Alzheimer's patients; and the seventh is caused by molecules like collagen becoming degraded through cross-linking. Cataracts and stiff joints are two age-related conditions that are caused by this last process. These seven types of damage include a large number of subtypes, each of which might require its own remedy. For example, recent research on breast cancer has found that it comprises ten separate diseases, each with its own characteristic genetic profile, response to treatment, and mortality rate.[17] How many separate fixes would be needed to "cure" aging, should this even be possible, is simply not known.

One of the big challenges is presented by a mechanism that

stops old cells from dividing. Some of these senescent cells die, but those that survive perpetrate a crime of omission and a crime of commission. The omission is that their inability to divide prevents them from helping to repair tissues. The crime of commission is the poisoning of the other cells around them. When this phenomenon of cellular senescence was discovered by Leonard Hayflick back in 1961,[18] the initial skepticism was followed by great excitement because it seemed like an obvious cause of aging.[19] Hayflick discovered that he could culture human cells quite successfully in the lab through 40–60 replications, but that after that point the cells ran out of steam and refused to divide. This point became known as the Hayflick limit, after its discoverer. Just what stopped the cells from dividing was a mystery, but whatever it was seemed like a ticking clock that might place an upper limit on how long a person could live in a good state of health.

The identity of the clock and how it works were gradually revealed in the 1970s and 1980s. It turned out to be a structure involved with the replication of DNA, a process that takes place every time a cell divides.[20] The DNA molecules in a human cell are extremely thin and long. Stretched out in a line, the DNA in a single cell would be between six and nine feet long.[21] Packing such molecules into a tiny cell is a feat of natural nanoengineering to be marveled at. The packages of hyper-coiled DNA in cells are called chromosomes, and each human cell has 23 pairs of these.

The process that copies the DNA in a chromosome has a problem when it gets to the end of the molecule, where it tends to stop short, leaving loose ends like the unraveling sleeves of an old sweater. This problem was fixed very early in the evolution of the eukaryotes by the placement of a cap, called a telomere, at either end of each chromosome. Elizabeth Blackburn and her collaborators, working at Yale and later at the University of California, Berkeley, discovered the structure of telomeres, which turned out to be made of a repeating DNA sequence of six bases.

The telomeres do not keep a chromosome from getting shorter at the ends each time it is replicated, but they prevent the genes in the chromosome from being clipped by taking the hit for them. Each time a cell divides, the telomeres on the chromosomes of its daughter cells get shorter. Of course the telomeres eventually get clipped to a nub, and at that point the cells lose the ability to divide and enter a state called replicative senescence.

Imagine that the problem of the loose ends left by DNA replication was presented to you as an engineering challenge and that there was a prize of eternal youth if you could solve it. You might, if you were as clever as 2 billion years of evolution, come up with the telomere as a solution. You'd proudly present your solution to the Nobel Prize committee in heaven, and they'd say, "Hang on a minute! What are you going to do about cells in the germ line? Your egg and sperm cells are going to stop dividing when their telomeres hit the buffers, just like all the rest of the cells." No immortality for you!

There must, of course, be a solution, and it was discovered in 1985 by Carol Greider, a graduate student of Elizabeth Blackburn. Greider discovered an enzyme called telomerase that repairs the telomeres in germ line cells, restoring them to their original length during DNA replication. In 2009 Elizabeth Blackburn, Carol Greider, and Jack Szostak shared the Nobel Prize in Physiology or Medicine for working out the telomere story. So now we know that the telomere clock counts down cell replications and that the enzyme telomerase keeps the clock wound up in germ line cells. Does this story explain the cause of aging and solve the problem of limited life spans? For a while it seemed a good bet that it did, suggesting to Hayflick and others that replicative senescence limits life span.[22] Maybe G. C. Williams was wrong, for once, and there's an elixir of life in a bottle labeled "Telomerase" available in your local pharmacy. But don't count on it.

The problem—as only Douglas Adams, author of *The Hitch-*

hiker's Guide to the Galaxy, could have guessed—is the mice. In *The Hitchhiker's Guide*, Earth turns out to be a planet-sized computer designed by mice, who have constructed it to find the mysterious question to which the answer is 42. Well, while they were at it, the mice also corrected us on the replicative senescence hypothesis. Mouse cell lines are immortal, though mice themselves, last time I looked it up (see the appendix), are not. Given oxygen and fresh nutrients, mouse cells can replicate indefinitely in the lab; they show no Hayflick limit because their somatic cells contain telomerase and they have telomeres that are up to ten times as long as those in human cells.[23] Whatever it is that limits mouse lifetimes to four years, it cannot be replicative senescence, and if replicative senescence doesn't limit life span in mice, why should it do so in other species?

So here is yet another example of a hypothesis, like the rate of living and oxidative stress hypotheses, that at first sight seems to offer an obvious, general solution to the entire problem of why organisms senesce, but breaks down when species are compared. G. C. Williams must be chuckling in his grave, and on his headstone we should expect to read the words "I told you so."

But from the ashes of every burned-out hypothesis arise the green shoots of a new one. We now need an explanation as to why mouse somatic cells contain telomerase while human somatic cells do not. Here are a couple of clues: First, all cancer cells produce telomerase. Second, if you add telomerase to human cells in culture, the Hayflick limit disappears and they can replicate indefinitely.[24] A hypothesis that is suggested by these clues is that the absence of telomerase in humans is an adaptation that lowers the risk of cancer. Recall Peto's paradox in chapter 2: mice and humans have similar rates of cancer, even though there are vastly more cells and many more cell replication events in a human lifetime than in the short life span of a mouse. We can deduce from this that there must be some very good brakes on runaway cell division in bigger, longer-lived animals. Is switching off the production of telomerase in somatic cells one of those brakes? It seems almost certain that it is.

A comparison of telomerase activity in fifteen different rodent species found that it varies a great deal among species and that the variation correlates with body size, but not with life span.[25] For instance, the eastern gray squirrel and the American beaver have similar maximum life spans of 24 and 23 years, respectively, but the beaver, which is forty times the weight of the gray squirrel, has only 13 percent of the telomerase activity of the smaller animal. It seems that the elevated risk of cancer that goes with large body size is counterbalanced by a reduction in telomerase activity. Natural selection seems to have made this adjustment many times independently in different mammal lineages. The critical size at which telomerase becomes a costly cancer risk and is switched off almost entirely in somatic cells is only about two pounds.[26]

The length of telomeres varies between species, but not in the way that would be expected if their length determined how long a species could typically live. In the absence of telomerase, telomeres become shorter with each cell division until they become so short that the cell hits the Hayflick limit and stops dividing. The longer a telomere is to start with, the more cell divisions it takes to reach the limit at which replicative senescence occurs. Therefore, if the onset of replicative senescence limited the life span, you would expect long-lived species to have longer telomeres than short-lived species. In fact, the opposite pattern is seen: telomere length is inversely correlated with life span among mammals, and long-lived mammals such as our own species have short telomeres. These observations suggest that short telomeres evolved because they act as a further brake that prevents cancer from emerging in long-lived species. Of course this brake will work only in species in which natural selection has already switched off telomerase, because telomerase prevents telomeres from becoming shorter at each cell division.

So replicative senescence caused by short telomeres might have a role in aging in longer-lived species after all. If it does, then it would be an example of the double-acting mutations predicted by the evolutionary theory of senescence. Replicative senescence is the downside experienced in later life of mechanisms that prevent cancer during youth. The idea that short telomeres exact such a penalty has been put to the test in several wild bird species, with remarkably consistent results. In alpine swifts, American tree swallows, European jackdaws, and southern giant petrels, individuals with longer telomeres in the chromosomes of their red blood cells had higher survival rates than those with shorter telomeres.[27] A similar relationship has been observed between mortality and telomere length in the white blood cells of people aged 60 or more in Utah.[28] Compared with people who had long telomeres, those who had short ones experienced three times the mortality rate from heart disease and more than eight times the mortality rate from infection.

The association of telomere length with survival may be direct, indirect, or both. For example, short telomeres could have a direct effect on susceptibility to infection if they handicap the rate at which new white blood cells, whose job it is to fight infection, are generated by cell division. Telomere length might equally well be an indirect marker of other aging processes such as oxidative stress. Telomere replication is known to be more sensitive to oxidative stress than replication of other parts of the chromosome, and this sensitivity may cause telomeres to shorten.

Since the original Utah study was published in 2003, there have been many thousands of similar studies, though a review of those studies conducted in 2011 found that only a tiny fraction were good enough to draw firm conclusions from.[29] Just ten studies of human mortality passed muster, and of these, half showed a correlation between telomere length and survival, and half did not. Even though the evidence for birds looks promising, in humans there are probably too many influences on telomere length to make it a useful biomarker for aging. These influences include such things as how old your parents were when you were born, your state of health, whether you smoke, take multivitamins, or drink alcohol, and your socioeconomic status, body mass index, gender, and racial group. In view of the fact that another study has found that among the old, how good you look for your age is a predictor of mortality,[30] it's unlikely that there is any message hidden in the length of your telomeres that your best friend couldn't give you over a candid cup of coffee.

Whether or not telomere length is a predictor of health and mortality—which might depend on whether you are a mouse, a man, or a mynah bird—you are undoubtedly better off without a sluggish collection of senescent cells in your tissues. It has recently been shown, however, that in mice that have been suitably genetically engineered, senescent cells can be selectively removed by targeting them with a drug. In these mice, removing senescent cells not only slowed down aging processes in fat, muscle, and eye tissue, but even reversed damage that had already taken place.[31] Just as remarkably, another study induced senescent human cells to divide and to produce stem cells that possess restored telomere length and are also free of the thousand natural shocks that the flesh is heir to.[32] Do studies such as these herald the eventual abolition of senescence that Aubrey de Grey dreams of? Not just yet. You should only apply for the drug treatment that purges senescent cells if you are a mouse that had the foresight to be suitably prepared through genetic engineering while still a fertilized egg. Using senescent cells to produce stem cells might one day help with tissue repair in old age, but there is a long way to go on that yet.

De Grey's SENS agenda sounds like science fiction, but who knows what the future holds? The long-lived Howard families in *Methuselah's Children* leave Earth by spaceship to escape persecution by the ephemerals. After adventures on another planet, some of the Howards decide it would be better to return to Earth where they belong than to stay in an alien world. When they get back to Earth nearly 75 years after they left, they discover that there has been remarkable progress while they were gone, and that the ephemerals have invented a technology that prolongs human life.

In chapter 1, I promised that I would lay before your feet a mosaic of the modern scientific understanding of aging and longevity, not just in humans but also in plants and animals. Now let me bring all the pieces together and show you how they fit into as grand a pattern as the one in the medieval Great Pavement of Westminster Abbey. The pattern emerges from a panoply of paradoxes. Everything about aging and longevity starts as a puzzle. When you first opened this book, you probably had somewhere in your mind a question like, "Why don't we live longer than we do?" For some this question is an obsession that ends in the freezers of the Alcor Life Extension Foundation, where, sadly for the sales of this book, there is no reading after lights out.

Historically speaking, the question is the wrong way around, because life on Earth began as tiny ephemeral beings and continued that way for 2 billion years. The first step in life extension was from single-celled organisms to the coalitions of cells that form multicellular organisms capable of self-replacement and repair. Ironically, it was not until life became complex and long-lived that it could fret about its own brevity.

Next is the paradox that the forces of evolution that raised us from the mire seem indifferent to, or powerless to prevent, the aging and death that return us thence. "Ashes to ashes, dust to dust" has an appealing symmetry about it that may satisfy the poet, but not the practical scientist, who can think of better ways of doing things. Why does natural selection, with its blind, single-minded focus on traits that endure, let organisms that have made it successfully through the perils of life just decay and slip away? The answer eluded science for nearly a century after Charles Darwin discovered natural selection. Then Peter Medawar and a few others realized that the answer lay in the diminishing contribution that individuals make to future generations as they grow older. Natural selection retires in old age, and this has permitted the accumulation of mutations that damage cells and interfere with body maintenance in later life. More diabolically yet, selection actually favors mutations that cause senescence if these same mutations benefit reproduction during earlier life.

There are just two escape clauses in the iron rule that natural selection tolerates the ravages of senescence. These escape clauses are not exceptions to the rules of natural selection, but rather accommodations to it. The first one covers organisms that increase the number of offspring they bear as they get older. Some fishes, lobsters, and giant clams, as well as many plants, can do so because they continue to grow larger with increasing age, which permits them to reproduce more and more. The organisms in question can all live for at least a century. In the case of plants, some live for millennia because they benefit from the second escape clause as well.

In most animals the germ line cells that produce sperm and eggs are anatomically separated from the other cells of the body, or the soma. The separation of germ line and soma allows natural selection to abandon the maintenance of the soma in later life without damaging the germ line. However, plants and colonial animals do not separate germ line and soma, and as a consequence, natural selection continues to defend them against the damaging effects of mutation as they get older. As a consequence, these organisms can be very, very long-lived, though many do senesce, and some are very short lived.

Plants are protected by the unity of germ line and soma from mutation accumulation and the double-acting mutations that cause senescence in animals. It therefore seems odd, if not to say ungrateful, that organisms carrying a "get out of aging free" card often choose not to use it, but instead bloom and die as quickly as a poppy. The explanation is in the environments where such short-lived plants live. If conditions are such that survival from year to year is low and uncertain, natural selection will favor early, copious reproduction and hang the consequences. Reproduction invariably has a cost, and in extreme cases that cost may be an early death. Pacific salmon know all about that.

The extrinsic risk of mortality to which adults are exposed affects whether natural selection favors a life that is short or one that is long. The metaphor of a train with carriages coupled by fragile links explains rather neatly why flying animals, animals that live in burrows, and animals that are protected from predators by poison or body armor live longer than animals that lack these features. However, an explanation at the cellular level for why some species senesce faster than others has proved much more elusive. One likely sounding hypothesis after another has been proposed, only to be found wanting in its generality when weighed against all the evidence. Yet there is an evolutionary explanation for there being no single cause of aging. The reason is that at the point in the life span where natural selection loses its power altogether, anything goes, and everything does. Before that point natural selection fixes the weakest links, ensuring that cellular function is not vulnerable to failed maintenance.

I have saved the strangest paradox for last, as it is arguably the most important one from a practical point of view, though it is often forgotten. It is this: despite the fact that senescence has not been conquered in humans, a huge increase in average human longevity has taken place since 1840, with 15 minutes added to the life span every hour over the last 170 years. Much of this increase is due to falling infant mortality rates, though improvements in adult health have also contributed to it. These measures have postponed senescence, not defeated it. If so much progress has been made without fundamentally altering senescence, we need to ask ourselves whether further improvements in life span will be more attainable through a program like SENS or through improvements to health in old age.

Life span is greater on average in richer countries than in poorer ones, but the relationship between wealth and life span is not linear. Data from the United Nations Development Programme show that as personal income rises, from almost nothing in the poorest African countries toward $10,000 a year in, say, Turkey, there is a steep rise in life expectancy from 40 years of age to around 70. Beyond that, each additional $10,000 a year buys a smaller and smaller improvement in average life span. The reason is not only that further improvement is increasingly difficult and expensive, but also that another economic factor comes into play: inequality of incomes among people within the population.[33]

Within the United States, where the gap in income between the richest and the poorest citizens differs among the fifty states of the union, life expectancy tends to be highest in those states where the income gap is least. The same trend is seen among countries. The gap between rich and poor is smallest in Japan, where life expectancy is also greatest. Sweden falls a short way behind Japan on both measures, while Portugal, the United States, and Singapore have the highest income inequalities and the lowest life expectancies among developed nations. The notable thing about these trends is that they are independent of wealth per se. Income per capita in Portugal is half that in the United States, but the gap between rich and poor is large in both countries and accounts for their equally poor performance in life expectancy.[34]

Why income inequality should influence life span in this way in developed countries is a complicated matter with political, economic, social-psychological, and biological causes. If there is good news in this unexpected finding, it is that you do not need to be a biologist to do something about it. And that, dear reader, is the long and the short of it.

APPENDIX

Scientific Names of Species Mentioned in the Text

This appendix gives the scientific names of species mentioned in the text and their typical and/or maximum (in parentheses) known life spans in years. Animal data are mainly from the AnAge database (http://genomics.senescence.info/species/), and other data are mainly from sources cited in the notes for the relevant chapters. A dash indicates that life span is unknown.

COMMON NAME	SCIENTIFIC NAME	LIFE SPAN
African grey parrot	*Psittacus erithacus*	50
Alpine swift	*Apus melba*	6 (26)
Amazonian tree	*Cariniana micrantha*	1,400
American beaver	*Castor canadensis*	23
American eel	*Anguilla rostrata*	15 (50)
American robin	*Turdus migratorius*	17

COMMON NAME	SCIENTIFIC NAME	LIFE SPAN
Atlantic salmon	*Salmo salar*	13
autumn gentian	*Gentianella amarella*	0.25–1.5
balsam fir	*Abies balsamea*	> 80
bamboo	Bambusoideae	(120)
barn swallow	*Hirundo rustica*	16
birch	*Betula* spp.	100–200
black bear	*Ursus americanus*	(34)
bowhead whale	*Balaena mysticetus*	(211)
bracken fern (clones)	*Pteridium aquilinum*	(700)
bristlecone pine	*Pinus longaeva*	(4,789)
brown antechinus	*Antechinus stuartii*	1 (5.4)
burdock	*Arctium minus*	2
capelin	*Mallotus villosus*	10
capybara	*Hydrochaeris hydrochaeris*	10 (15)
century plant	*Agave americana*	25
chickweed	*Stellaria media*	< 1
common yellowthroat	*Geothlypis trichas*	(11.5)
crab spider	Lysiteles coronatus	—
creosote bush	*Larrea tridentata*	(ca. 11,000)
dibbler	*Parantechinus apicalis*	> 3 (5.5)
dodo	*Raphus cucullatus*	—
early gentian	*Gentianella anglica*	0.3
eastern gray squirrel	*Sciurus carolinensis*	(24)
eastern red cedar	*Juniperus virginiana*	(300)
eastern white cedar	*Thuja occidentalis*	80 (1,800)
emu	*Dromaius novaehollandiae*	16.6
European eel	*Anguilla anguilla*	10–15 (88)
evening primroses	*Oenothera* spp.	2–3

COMMON NAME	SCIENTIFIC NAME	LIFE SPAN
flamingo, greater	*Phoenicopterus roseus*	(44)
foxglove	*Digitalis purpurea*	2
freshwater pearl mussel	*Margaritifera margaritifera*	(250)
fruit fly	*Drosophila melanogaster* and other species	0.3
geoduck clam	*Panopea generosa* (syn. *P. abrupta*)	(169)
giant lobelia	*Lobelia telekii*	40–70
groundsel	*Senecio vulgaris*	< 1
gwarrie tree	*Euclera undulata*	(ca. 10,000?)
hard clam	*Mercenaria mercenaria*	68 (106)
herring gull	*Larus argentatus*	(49)
honeybee queen	*Apis mellifera*	(8)
honeybee worker	*Apis mellifera*	< 1
house mouse	*Mus musculus*	(4)
human	*Homo sapiens*	66 (122)
human fish or olm	*Proteus anguinus*	(100)
jackdaw	*Corvus monedula*	(20)
Japanese hump earwig	*Anechura harmandi*	1
killer whale	*Orcinus orca*	50 (100)
kosi palm	*Raphia australis*	30
lapwing	*Vanellus vanellus*	(16)
long-leaved plantain	*Plantago lanceolata*	1–2
Madagascan chameleon	*Furcifer labordi*	0.4
Malaysian treehopper	*Pyrgauchenia tristaniopsis*	0.2
medfly	*Ceratitis capitata*	0.1
Mexican *Astrocaryum* palm	*Astrocaryum mexicanum*	123
mullein	*Verbascum thapsus*	2

COMMON NAME	SCIENTIFIC NAME	LIFE SPAN
naked mole-rat	*Heterocephalus glaber*	25 (31)
nematode worm*	*Caenorhabditis elegans*	0.06
ocean quahog	*Arctica islandica*	100 (405)
ostrich	*Struthio camelus*	(50)
Outeniqua yellowwood	*Afrocarpus falcatus*	(650)
Pacific salmon—coho	*Oncorhynchus kisutch*	3
periodical cicadas	*Magicicada* spp.	13; 17
pipistrelle bat	*Pipistrellus pipistrellus*	(16)
ponderosa pine	*Pinus ponderosa*	300
poppy	*Papaver* spp.	< 1
puya	*Puya raimondii*	80–150
rat	*Rattus norvegicus*	3.8
southern giant petrel	*Macronectes giganteus*	(40)
spear thistle	*Cirsium vulgare*	2
stout infantfish	*Schindleria brevipinguis*	0.16
sulfur pearl bacterium	*Thiomargarita namibiensis*	—
talipot palm	*Corypha umbraculifera*	30–80
Tasmanian devil	*Sarcophilus harrisii*	2
thale cress	*Arabidopsis thaliana*	0.12
tree swallow	*Tachycineta bicolor*	(12)
trembling aspen (clones)	*Populus tremuloides*	(10,000)
tuberculosis bacterium	*Mycobacterium tuberculosis*	—
ulcer-causing bacterium	*Helicobacter pylori*	—
Virginia opossum	*Didelphis virginiana*	2–3 (6.5)
western red cedar	*Thuja plicata*	> 1,000
wild carrot	*Daucus carota*	2–3
wild strawberry	*Fragaria vesca*	3–10
willow	*Salix* spp.	55 (85)

COMMON NAME	SCIENTIFIC NAME	LIFE SPAN
Wollemi pine	*Wollemia nobilis*	> 350
Yew	*Taxus baccata*	> 1,000

*There are tens of thousands of known species of worms in the phylum Nematoda, and certainly a much larger number of undescribed species, but here I use the common name to refer only to this one species.

NOTES

Chapter One

1. *Night is the morning's Canvas:* E. Dickinson, *The Complete Poems of Emily Dickinson*, ed. T. H. Johnson (Little Brown, 1960), 9.

2. *An inscription in Latin:* R. Foster, *Patterns of Thought: The Hidden Meaning of the Great Pavement of Westminster Abbey* (Jonathan Cape, 1991), 3.

3. *the jawbone of King Richard II:* R. Jenkyns, *Westminster Abbey*, Wonders of the World (Profile, 2006), 216.

4. *"I did kiss a Queen":* S. Pepys, *The Diary of Samuel Pepys* (vol. 3, p. 357, February 23, 1669), ed. J. Warrington (Dent Dutton, 1953), 521.

5. *"What, thought I, is this vast assemblage":* W. Irving, *The Sketch Book of Geoffrey Crayon, Gent.* (New American Library, 1961), 177–78.

6. *the memorial to William Congreve:* T. Trowles, *Westminster Abbey Official Guide* (Dean and Chapter of Westminster, 2005); C. Y. Ferdinand and D. F. McKenzie, "Congreve, William (1670–1729)," *Oxford Dictionary of National Biography*, ed. L. Goldman et al. (Oxford University Press, 2004), doi:10.1093/ref:odnb/6069.

7. *The pinnacle of pomp was reached:* Jenkyns, *Westminster Abbey*, 169.

8. *he laughed only once:* J. Holt, *Stop Me if You've Heard This: A History and Philosophy of Jokes* (Profile Books, 2008), 62–63.

9. *Annie died of tuberculosis:* R. Keynes, *Annie's Box* (Fourth Estate, 2001).

10. *"endless forms most beautiful":* C. Darwin, *The Origin of Species by Means of Natural Selection*, 1st ed. (1859; reprint, Penguin, 1968).

11. *a tuberculosis ward: Wikipedia*, s.v. "List of tuberculosis cases," accessed March 26, 2011, http://en.wikipedia.org/wiki/List_of_tuberculosis_cases.

12. *evolutionary mark on the human genome:* M. Moller, E. de Wit, and E. G. Hoal, "Past, present and future directions in human genetic susceptibility to tuberculosis," *FEMS Immunology & Medical Microbiology* 58 (2010): 3–26.

13. *epidemics of the past:* F. O. Vannberg, S. J. Chapman, and A. V. S. Hill, "Human genetic susceptibility to intracellular pathogens," *Immunological Reviews* 240 (2011): 105–16.

14. *Death in childbirth: Wikipedia*, s.v. "List of women who died in childbirth: United Kingdom, accessed March 26, 2011, http://en.wikipedia.org/wiki/List_of_women_who_died_in_childbirth#United_Kingdom.

15. *the bacterium* Helicobacter pylori: G. Morelli et al., "Microevolution of *Helicobacter pylori* during prolonged infection of single hosts and within families," *PLoS Genetics* 6 (2010), doi:10.1371/journal.pgen.1001036.

16. *jumped from humans to big cats:* M. Eppinger et al., "Who ate whom? Adaptive *Helicobacter* genomic changes that accompanied a host jump from early humans to large felines," *PLoS Genetics* 2 (2006): e120.

Chapter Two

1. *"And what is Life?":* J. Clare, *Poems Chiefly from Manuscript*, ed. E. Blunden and A. Porter (Cobden-Sanderson, 1920).

2. *nearly every organism was single-celled:* R. K. Grosberg and R. R. Strathmann, "The evolution of multicellularity: A minor major transition?," *Annual Review of Ecology, Evolution, and Systematics* 38 (2007): 621–54, doi:10.1146/annurev.ecolsys.36.102403.114735.

3. *outnumbered at least ten to one:* M. Wilson, *Bacteriology of Humans: An Ecological Perspective* (Blackwell, 2008).

4. *"I contain multitudes":* W. Whitman, *Leaves of Grass* (Airmont Publishing, 1965), 79, sect. 51.

5. *in rocks buried nearly two miles:* D. Chivian et al., "Environmental genomics reveals a single-species ecosystem deep within Earth," *Science* 322, no. 5899 (2008): 275–78, doi:10.1126/science.1155495.

6. *no human could survive:* F. Bäckhed et al., "Host-bacterial mutualism in the human intestine," *Science* 307, no. 5717 (2005): 1915–20, doi:10.1126/science.1104816.

7. *The biggest bacterium known is the sulfur pearl:* H. N. Schulz et al., "Dense populations of a giant sulfur bacterium in Namibian shelf sediments," *Science* 284, no. 5413 (1999): 493–95, doi:10.1126/science.284.5413.493.

8. *a new species:* M. D. Vincent, "The animal within: Carcinogenesis and the clonal evolution of cancer cells are speciation events *sensu stricto*," *Evolution* 64, no. 4 (2010): 1173–83, doi:10.1111/j.1558-5646.2009.00942.x.

9. *their own best-selling biography:* R. Skloot, *The Immortal Life of Henrietta Lacks* (Macmillan, 2010).

10. *a venereal disease in dogs:* A. M. Leroi et al., "Cancer selection," *Nature Reviews Cancer* 3, no. 3 (2003): 226–31.

11. *tumors on different animals:* A. M. Pearse and K. Swift, "Allograft theory: Transmission of devil facial-tumour disease," *Nature* 439, no. 7076 (2006): 549.

12. *now listed as endangered:* C. E. Hawkins et al., "Emerging disease and population decline of an island endemic, the Tasmanian devil *Sarcophilus harrisii*," *Biological Conservation* 131, no. 2 (2006): 307–24.

13. *cells in the surface lining of your gut:* S. A. Frank and M. A. Nowak, "Problems of somatic mutation and cancer," *Bioessays* 26, no. 3 (2004): 291–99, doi:10.1002/bies.20000.

14. *colorectal cancer in 90-year-old humans:* A. F. Caulin and C. C. Maley, "Peto's Paradox: Evolution's prescription for cancer prevention," *Trends in Ecology & Evolution* 26, no. 4 (2011): 175–82.

15. *Richard Peto observed:* R. Peto et al., "Cancer and ageing in mice and men," *British Journal of Cancer* 32, no. 4 (1975): 411–26.

16. *bigger species are better protected:* J. D. Nagy et al., "Why don't all whales have cancer? A novel hypothesis resolving Peto's paradox," *Integrative and Comparative Biology* 47, no. 2 (2007): 317–28, doi:10.1093/icb/icm062.

17. *the record for vertebrate longevity:* S. N. Austad, "Methusaleh's zoo: How nature provides us with clues for extending human health span," *Journal of Comparative Pathology* 142 (2010): S10–S21.

18. *genes protecting us from cancer:* A. Budovsky et al., "Common gene signature of cancer and longevity," *Mechanisms of Ageing and Development* 130, no. 1–2 (2009): 33–39, doi:10.1016/j.mad.2008.04.002; R. Tacutu et al., "Molecular links between cellular senescence, longevity and age-related diseases: A systems biology perspective," *Aging* 3, no. 12 (2011): 1178–91.

19. *bivalves are some of the longest-lived animals:* I. D. Ridgway et al.,

"Maximum shell size, growth rate, and maturation age correlate with longevity in bivalve molluscs," *Journals of Gerontology, Series A, Biological Sciences and Medical Sciences* 66, no. 2 (2011): 183–90, doi:10.1093/gerona/glq172.

20. *the stout infantfish:* W. Watson and H. J. Walker, "The world's smallest vertebrate, *Schindleria brevipinguis*, a new paedomorphic species in the family Schindleriidae (Perciformes: Gobioidei)," *Records of the Australian Museum* 56, no. 2 (2004): 139–42.

21. *bigger species do indeed live longer than smaller ones:* J. P. de Magalhães et al., "An analysis of the relationship between metabolism, developmental schedules, and longevity using phylogenetic independent contrasts," *Journals of Gerontology, Series A, Biological Sciences and Medical Sciences* 62, no. 2 (2007): 149–60.

22. *The common pipistrelle bat:* J. P. de Magalhães and J. Costa, "A database of vertebrate longevity records and their relation to other life-history traits," *Journal of Evolutionary Biology* 22, no. 8 (2009): 1770–74, doi:10.1111/j.1420-9101.2009.01783.x.

23. *Naked mole-rats:* R. Buffenstein, "The naked mole-rat: A new long-living model for human aging research," *Journals of Gerontology, Series A, Biological Sciences and Medical Sciences* 60, no. 11 (2005): 1369–77.

24. *Birds, like bats, have unusually long lives:* J. P. de Magalhães et al., "An analysis of the relationship between metabolism, developmental schedules, and longevity using phylogenetic independent contrasts," *Journals of Gerontology, Series A, Biological Sciences and Medical Sciences* 62, no. 2 (2007): 149–60.

25. *flamingos and their relatives are the longest-lived birds:* D. E. Wasser and P. W. Sherman, "Avian longevities and their interpretation under evolutionary theories of senescence," *Journal of Zoology* 280, no. 2 (2010): 103–55, doi:10.1111/j.1469-7998.2009.00671.x.

26. *crows have been known to fashion tools:* A. Seed and R. Byrne, "Animal tool-use," *Current Biology* 20, no. 23 (2010): R1032–R1039, doi:10.1016/j.cub.2010.09.042.

27. *other exceptional inhabitants of Methuselah's menagerie:* Austad, "Methusaleh's zoo."

28. *oldest person buried in Westminster Abbey:* K. Thomas, "Parr, Thomas (d. 1635), supposed centenarian," *Oxford Dictionary of National Biography*, ed. L. Goldman et al. (Oxford University Press, 2004), doi:10.1093/ref:odnb/21403.

29. *a poet named John Taylor:* J. Taylor, *The Old, Old, Very Old Man*, 1635, accessed December 27, 2010, http://www.archive.org/details/oldoldveryoldman00tayliala.

30. *Francis Bacon (1561–1626):* D. B. Haycock, *Mortal Coil: A Short History of Living Longer* (Yale University Press, 2008).

31. *an ingenious, if flawed, explanation:* Haycock, *Mortal Coil*, 23.

32. *"In those green-pastured mountains of Forta-fe-Zee":* Dr. Seuss, *You Are Only Old Once: A Book for Obsolete Children* (Random House, 1986).

33. *Grace Halsell, author of the book* Los Viejos: G. Halsell, *Los Viejos: Secrets of Long Life from the Sacred Valley* (Rodale Press, 1976).

34. *claims of extreme old age in Vilcabamba:* R. B. Mazess and S. H. Forman, "Longevity and age exaggeration in Vilcabamba," *Journal of Gerontology* 34 (1979): 94–98.

35. *A study of life expectancy:* R. B. Mazess and R. W. Mathisen, "Lack of unusual longevity in Vilcabamba, Ecuador," *Human Biology* 54, no. 3 (1982): 517–24.

36. *one supposed Shangri-La after another:* R. D. Young et al., "Typologies of extreme longevity myths," *Current Gerontology and Geriatrics Research* (2011), doi:10.1155/2010/423087.

37. *Frenchwoman Jeanne Calment:* B. Jeune et al., "Jeanne Calment and her successors: Biographical notes on the longest living humans," in *Supercentenarians*, ed. H. Maier et al., Demographic Research Monographs (Springer, 2010).

38. *Dan Buettner, a journalist:* "Dan Buettner," Field Notes, *National Geographic*, accessed May 2, 2011, http://ngm.nationalgeographic.com/2005/11/longevity-secrets/buettner-field-notes.

39. *a comfortable record:* Y. Voituron et al., "Extreme lifespan of the human fish (*Proteus anguinus*): A challenge for ageing mechanisms," *Biology Letters* 7, no. 1 (2011): 105–7, doi:10.1098/rsbl.2010.0539.

Chapter Three

1. *"And after many a summer dies the swan":* Alfred, Lord Tennyson, "Tithonus" (1860), in *Poems of Tennyson* (Oxford University Press, 1918), 616.

2. *there was once a mortal by the name of Tithonus:* R. Graves, *Greek Myths* (Penguin, 1957).

3. *"Senescence begins":* O. Nash, *The Pocket Book of Ogden Nash* (Simon & Schuster, 1962).

4. *an American male aged 50:* Data from World Health Organization, accessed April 8, 2012,http://apps.who.int/gho/data/.

5. *aging is one of the leading causes of statistics:* L. Hayflick, *How and Why We Age* (Ballantine, 1994), 53.

6. *the sin of usury:* R. H. Tawney, *Religion and the Rise of Capitalism* (Penguin, 1926).

7. *Outram's woe-filled lamentation:* "The Annuity," by George Outram, in *Verse and Worse*, ed. A. Silcock (Faber & Faber, 1958).

8. *One of the investors:* C. Mitchell and C. Mitchell, "Wordsworth and the old men," *Journal of Legal History* 25, no. 1 (2004): 31–52.

9. *"Upon the forest-side in Grasmere Vale":* W. Wordsworth, "Michael: A Pastoral Poem" (1800), lines 40–47, in *The Poetical Works of Wordsworth*, ed. T. Hutchinson (Oxford University Press, 1932).

10. *In 1779, one Benjamin Gompertz:* D. P. Miller, "Gompertz, Benjamin (1779–1865)," *Oxford Dictionary of National Biography*, ed. L. Goldman et al. (Oxford University Press, 2004).

11. *The Mortality Rate Doubling Time:* C. E. Finch, *Longevity, Senescence and the Genome* (University of Chicago Press, 1990), 23.

12. *Two hundred years ago:* The change over the last 200 years is shown in a powerful animated graphic at www.gapminder.org (accessed July 10, 2011). Access the graphic via www.bit.ly/cVMWJ4.

13. *life expectancy has increased:* J. Oeppen and J. W. Vaupel, "Demography: Broken limits to life expectancy," *Science* 296, no. 5570 (2002): 1029–31.

14. *female life expectancy in Sweden was 83 years:* WolframAlpha, accessed July 9, 2011, http://www.wolframalpha.com. You can check the latest statistics for any country by entering a search term such as "life expectancy female USA" in the computational knowledge engine at WolframAlpha.com.

15. *remarkable advances in life expectancy:* Oeppen and Vaupel, "Demography."

16. *Life expectancy in the United States has increased:* WolframAlpha, accessed July 9, 2011, http://www.wolframalpha.com.

17. *countries where smoking is especially prevalent:* K. Christensen et al., "Ageing populations: The challenges ahead," *Lancet* 374, no. 9696 (2009): 1196–208.

18. *male life expectancy in Russia:* WolframAlpha, accessed July 10, 2011, http://www.wolframalpha.com/input/?i=male+life+expectancy+russia.

19. *the lapwing and the herring gull:* Finch, *Longevity, Senescence and the Genome*, 122.

20. *women in New Zealand:* Oeppen and Vaupel, "Demography."

21. *the majority of children born since 2000:* Christensen et al., "Ageing populations."

22. *satirical poem "Chard Whitlow":* H. Reed, "Chard Whitlow," *Statesman & Nation* 21, no. 533 (1941): 494.

23. *one-third of a Danish group of centenarians:* K. Christensen et al., "Exceptional longevity does not result in excessive levels of disability," *Proceedings of the National Academy of Sciences of the United States of America* 105, no. 36 (2008): 13274–79, doi:10.1073/pnas.0804931105.

24. *40 percent of a group of American supercentenarians:* Christensen et al., "Ageing populations."

25. *shorter-lived ancestors:* C. Selman and D. J. Withers, "Mammalian models of extended healthy life span," *Philosophical Transactions of the Royal Society B: Biological Sciences* 366, no. 1561 (2011): 99–107, doi:10.1098/rstb.2010.0243.

26. *the mortality rate in this group comes to a standstill:* J. Gampe, "Human mortality beyond age 110," in *Supercentenarians*, ed. H. Maier et al., Demographic Research Monographs (Springer, 2010).

27. *medfly-rearing facility in southern Mexico:* J. Hendrichs et al., "Medfly area wide sterile insect technique programmes for prevention, suppression or eradication: The importance of mating behavior studies," *Florida Entomologist* 85, no. 1 (2002): 1–13.

28. *it was another 82 days before the last fly died:* J. R. Carey, *Longevity: The Biology and Demography of Life Span* (Princeton University Press, 2003).

29. *Males live longer than females in rats:* S. N. Austad, "Why women live longer than men: Sex differences in longevity," *Gender Medicine* 3, no. 2 (2006): 79–92.

30. *the appearance of a declining mortality rate:* J. W. Vaupel and A. I. Yashin, "Heterogeneity's ruses: Some surprising effects of selection on population dynamics," *American Statistician* 39, no. 3 (1985): 176–85.

31. *"second childishness and mere oblivion":* W. Shakespeare, *As You Like It*, act 2, scene 7, in *Complete works of William Shakespeare*, RSC edition (Macmillan, 2006).

32. *senescence has not been reduced:* J. W. Vaupel, "Biodemography of human ageing," *Nature* 464, no. 7288 (2010): 536–42, doi:10.1038/nature08984.

Chapter Four

1. *in a book called* Over the Teacups: O. W. Holmes Sr., *Over the Teacups*, 1889, Kindle edition.

2. *genes account for between 25 and 35 percent:* C. E. Finch and R. E. Tanzi, "Genetics of aging," *Science* 278, no. 5337 (1997): 407–11, doi:10.1126/science.278.5337.407.

3. *A queen honeybee lives and reproduces for several years:* D. Munch et al.,

"Ageing in a eusocial insect: Molecular and physiological characteristics of life span plasticity in the honey bee," *Functional Ecology* 22, no. 3 (2008): 407–21, doi:10.1111/j.1365-2435.2008.01419.x.

4. *the rare form of Alzheimer's:* Finch and Tanzi, "Genetics of aging."

5. *twins born in Denmark, Finland, and Sweden:* J. V. Hjelmborg et al., "Genetic influence on human life span and longevity," *Human Genetics* 119, no. 3 (2006): 312–21, doi:10.1007/s00439-006-0144-y.

6. *A study at Leiden in Holland:* R. G. J. Westendorp et al., "Nonagenarian siblings and their offspring display lower risk of mortality and morbidity than sporadic nonagenarians: The Leiden Longevity Study," *Journal of the American Geriatrics Society* 57, no. 9 (2009): 1634–37, doi:10.1111/j.1532-5415.2009.02381.x.

7. *offspring also had lower mortality rates:* M. Schoenmaker et al., "Evidence of genetic enrichment for exceptional survival using a family approach: The Leiden Longevity Study," *European Journal of Human Genetics* 14, no. 1 (2005): 79–84.

8. *lower risks of heart attack:* Westendorp et al., "Nonagenarian siblings.

9. *Dauers have been found attached:* WormBook: The Online Review of *C. elegans* Biology, accessed July 24, 2011, http://www.wormbook.org/chapters/www_ecolCaenorhabditis/ecolCaenorhabditis.html.

10. *The Worm Breeder's Gazette: The Worm Breeder's Gazette*, accessed December 21, 2012, http://www.wormbook.org/wbg/.

11. *Unhampered by the need to find and court a mate:* W. A. Van Voorhies et al., "The longevity of *Caenorhabditis elegans* in soil," *Biology Letters* 1, no. 2 (2005): 247–49, doi:10.1098/rsbl.2004.0278.

12. *The first longevity gene:* D. B. Friedman and T. E. Johnson, "3 mutants that extend both mean and maximum life-span of the nematode, *Caenorhabditis elegans*, define the *age-1* gene," *Journals of Gerontology, Biological Sciences* 43, no. 4 (1988): B102–B109; D. B. Friedman and T. E. Johnson, "A mutation in the *age-1* gene in *Caenorhabditis elegans* lengthens life and reduces hermaphrodite fertility," *Genetics* 118, no. 1 (1988): 75–86.

13. *mainly due to a decrease in the rate of senescence:* T. E. Johnson, "Increased life-span of age-1 mutants in *Caenorhabditis elegans* and lower Gompertz rate of aging," *Science* 249, no. 4971 (1990): 908–12, doi:10.1126/science.2392681.

14. *"I left culture dishes with my almost-infertile mutants":* C. Kenyon, "The first long-lived mutants: Discovery of the insulin/IGF-1 pathway for ageing," *Philosophical Transactions of the Royal Society B: Biological Sciences* 366, no. 1561 (2011): 9–16, doi:10.1098/rstb.2010.0276.

15. *Mutation in the* daf-2 *gene:* C. Kenyon et al., "A *C. elegans* mutant that lives twice as long as wild-type," *Nature* 366, no. 6454 (1993): 461–64, doi:10.1038/366461a0.

16. *the worm version of the hormone insulin:* K. D. Kimura et al., "daf-2, an insulin receptor-like gene that regulates longevity and diapause in *Caenorhabditis elegans*," *Science* 277, no. 5328 (1997): 942–46, doi:10.1126/science.277.5328.942.

17. *also present in yeast, fruit flies, and mice:* M. Tatar et al., "The endocrine regulation of aging by insulin-like signals," *Science* 299, no. 5611 (2003): 1346–51.

18. *70 percent similar:* Kimura et al., "daf-2."

19. *mutant worms with disabled senses:* J. Apfeld and C. Kenyon, "Regulation of life span by sensory perception in *Caenorhabditis elegans*," *Nature* 402, no. 6763 (1999): 804–9.

20. *mutants are better protected:* A. Taguchi and M. F. White, "Insulin-like signaling, nutrient homeostasis, and life span," *Annual Review of Physiology* 70, no. 1 (2008): 191–212, doi:10.1146/annurev.physiol.70.113006.100533.

21. *a likely explanation:* E. Cohen and A. Dillin, "The insulin paradox: Aging, proteotoxicity and neurodegeneration," *Nature Reviews Neuroscience* 9, no. 10 (2008): 759–67, doi:10.1038/nrn2474.

22. *associated with longer life in humans and mice:* Y. Suh et al., "Functionally significant insulin-like growth factor I receptor mutations in centenarians," *Proceedings of the National Academy of Sciences of the United States of America* 105, no. 9 (2008): 3438–42, doi:10.1073/pnas.0705467105; Taguchi and White, "Insulin-like signaling."

23. *it controls the growth of cell size:* M. N. Hall, "mTOR—What does it do?," *Transplantation Proceedings* 40 (2008): S5–S8, doi:10.1016/j.transproceed.2008.10.009.

24. *increased their life span by about 10 percent:* D. E. Harrison et al., "Rapamycin fed late in life extends life span in genetically heterogeneous mice," *Nature* 460, no. 7253 (2009): 392–95.

25. *hundreds of different genes are associated with normal aging:* J. P. de Magalhães et al., "Genome-environment interactions that modulate aging: Powerful targets for drug discovery," *Pharmacological Reviews* 64, no. 1 (2012): 88–101, doi:10.1124/pr.110.004499.

26. *rapamycin can reverse defects in cells taken from progeria patients:* K. Cao et al., "Rapamycin reverses cellular phenotypes and enhances mutant protein clearance in Hutchinson-Gilford progeria syndrome cells," *Science Translational Medicine* 3, no. 89 (2011), doi:89ra5810.1126/scitranslmed.3002346.

27. *ameliorating some of the effects of normal aging on cells:* C. R. Burtner and B. K. Kennedy, "Progeria syndromes and ageing: What is the connection?," *Nature Reviews Molecular Cell Biology* 11, no. 8 (2010): 567–78, doi: 10.1038/nrm2944.

28. *People carrying two copies of ε4:* G. J. McKay et al., "Variations in apolipoprotein E frequency with age in a pooled analysis of a large group of older people," *American Journal of Epidemiology* 173, no. 12 (2011): 1357–64, doi: 10.1093/aje/kwr015.

29. *balances out the extra risk:* A. M. Kulminski et al., "Trade-off in the effects of the apolipoprotein E polymorphism on the ages at onset of CVD and cancer influences human life span," *Aging Cell* 10, no. 3 (2011): 533–41, doi: 10.1111/j.1474-9726.2011.00689.x.

30. *a tract called* Discorsi de la vita sobria: A. Cornaro, *Discourses on the Sober Life [Discorsi de la vita sobria]* (Thomas Y. Crowell, 1916), http://www.archive.org/details/discoursesonsobe00cornrich.

31. *between 1,500 and 1,700 calories a day:* G. Crister, *Eternity Soup: Inside the Quest to End Aging* (Harmony Books, 2010).

32. *a blurb by President George Washington:* Crister, *Eternity Soup.*

33. *People practicing it feel perpetually cold:* Crister, *Eternity Soup.*

34. *"You can live to 100":* Woody Allen, quoted in J. Lloyd and J. Mitchinson, *Advanced Banter: The QI Book of Quotations* (Faber & Faber, 2008), 8.

35. *two different studies with monkeys:* S. N. Austad, "Ageing: Mixed results for dieting monkeys," *Nature*, vol. advance online publication (2012), doi: 10.1038/nature11484.

36. *the usual suspects are often implicated:* Taguchi and White, "Insulin-like signaling"; L. Partridge et al., "Ageing in *Drosophila*: The role of the insulin/Igf and TOR signalling network," *Experimental Gerontology* 46, no. 5 (2011): 376–81, doi:10.1016/j.exger.2010.09.003; J. J. McElwee et al., "Evolutionary conservation of regulated longevity assurance mechanisms," *Genome Biology* 8, no. 7 (2007), doi:R13210.1186/gb-2007-8-7-r132.

Chapter Five

1. *"Show, in your words and images":* Dylan Thomas, *Collected Poems 1934–1952*, ed. W. Davies and R. Maud (Dent, 1994), 183.

2. *An even older one found in Nevada:* R. M. Lanner, *The Bristlecone Book: A Natural History of the World's Oldest Trees* (Mountain Press, 2007).

3. *eastern white cedars with 1,800 annual rings:* D. W. Larson, "The para-

dox of great longevity in a short-lived tree species," *Experimental Gerontology* 36, no. 4–6 (2001): 651–73.

4. *the oldest corals:* E. B. Roark et al., "Extreme longevity in proteinaceous deep-sea corals," *Proceedings of the National Academy of Sciences of the United States of America* 106, no. 13 (2009): 5204–8, doi:10.1073/pnas.0810875106.

5. *Hunter-gatherers can make it to 70:* M. Gurven and H. Kaplan, "Longevity among hunter-gatherers: A cross-cultural examination," *Population and Development Review* 33, no. 2 (2007): 321–65, doi:10.1111/j.1728-4457.2007.00171.x.

6. *pollen and seeds produced by the ancient trees:* R. M. Lanner and K. F. Connor, "Does bristlecone pine senesce?," *Experimental Gerontology* 36, no. 4–6 (2001): 675–85.

7. *now growing faster:* M. W. Salzer et al., "Recent unprecedented tree-ring growth in bristlecone pine at the highest elevations and possible causes," *Proceedings of the National Academy of Sciences of the United States of America* 106, no. 48 (2009): 20348–53, doi:10.1073/pnas.0903029106.

8. *there are only 627 species:* A. Farjon, *A Natural History of Conifers* (Timber Press, 2008).

9. *about 60,000 are trees:* C. Tudge, *The Secret Life of Trees* (Allen Lane, 2005), 30.

10. *many tropical trees have now been aged:* D. M. A. Rozendaal and P. A. Zuidema, "Dendroecology in the tropics: A review," *Trees—Structure and Function* 25, no. 1 (2011): 3–16, doi:10.1007/s00468-010-0480-3.

11. *a study of trees felled in a logging concession:* J. Q. Chambers et al., "Ancient trees in Amazonia," *Nature* 391, no. 6663 (1998): 135–36, doi:10.1038/34325.

12. *tropical forests are known to be highly dynamic:* M. Martinez-Ramos and E. R. Alvarez-Buylla, "How old are tropical rain forest trees?," *Trends in Plant Science* 3, no. 10 (1998): 400–405, doi:10.1016/s1360-1385(98)01313-2.

13. *some millenarian trees in the Amazon:* W. F. Laurance et al., "Inferred longevity of Amazonian rainforest trees based on a long-term demographic study," *Forest Ecology and Management* 190, no. 2–3 (2004): 131–43; R. Condit et al., "Mortality-rates of 205 Neotropical tree and shrub species and the impact of a severe drought," *Ecological Monographs* 65 (1995): 419–39.

14. *the oldest trees are the slowest-growing ones:* S. Vieira et al., "Slow growth rates of Amazonian trees: Consequences for carbon cycling," *Proceedings of the National Academy of Sciences of the United States of America* 102, no. 51 (2005): 18502–7, doi:10.1073/pnas.0505966102.

15. *such as the Mexican* Astrocaryum *palm:* J. Silvertown et al., "Evolution

of senescence in iteroparous perennial plants," *Evolutionary Ecology Research* 3 (2001): 1–20.

16. *I studied another clear example myself in the Adirondacks:* J. Silvertown, *Demons in Eden: The Paradox of Plant Diversity* (University of Chicago Press, 2005).

17. *old shoots grow with the same vigor as young ones:* M. Mencuccini et al., "Evidence for age- and size-mediated controls of tree growth from grafting studies," *Tree Physiology* 27, no. 3 (2007): 463–73.

18. *quotes a nonsense poem:* J. Joyce, *A Portrait of the Artist as a Young Man* (Penguin 1965), chap. 1.

19. *harder for a single mutant plant cell to multiply out of control:* J. H. Doonan and R. Sablowski, "Walls around tumours—why plants do not develop cancer," *Nature Reviews Cancer* 10, no. 11 (2010): 793–802, doi:10.1038/nrc2942.

20. *apple and flower varieties originated in this way:* N. Kingsbury, *Hybrid: The History and Science of Plant Breeding* (University of Chicago Press, 2009).

21. *mutational variation of this kind is surprisingly rare:* E. J. Klekowski Jr., *Mutation, Developmental Selection, and Plant Evolution* (Columbia University Press, 1988).

22. *"A year for the stake. Three years for the field":* R. Foster, *Patterns of Thought: The Hidden Meaning of the Great Pavement of Westminster Abbey* (Jonathan Cape, 1991): 101.

23. *"There is a Yew-tree, pride of Lorton Vale":* William Wordsworth, "Yew Trees," in *Wordsworth's Poetical Works*, Oxford Edition (Oxford University Press, 1932), 84.

24. *still survives in Lorton:* "'Yew Trees' by William Wordsworth," Visit Cumbria, accessed September 12, 2012, http://www.visitcumbria.com/cm/lorton-yew-trees.htm.

25. *Trees with dense wood:* J. Chave et al., "Towards a worldwide wood economics spectrum," *Ecology Letters* 12, no. 4 (2009): 351–66, doi:10.1111/j.1461-0248.2009.01285.x.

26. *as much as 86 percent resin by weight:* C. Loehle, "Tree life histories: The role of defences," *Canadian Journal of Forest Research* 18 (1988): 209–22.

27. *chemically defended species live longer:* M. A. Blanco and P. W. Sherman, "Maximum longevities of chemically protected and non-protected fishes, reptiles, and amphibians support evolutionary hypotheses of aging," *Mechanisms of Ageing and Development* 126, no. 6–7 (2005): 794–803, doi:10.1016/j.mad.2005.02.006.

28. *Several studies of tree rings:* S. E. Johnson and M. D. Abrams, "Age class, longevity and growth rate relationships: Protracted growth increases in

old trees in the eastern United States," *Tree Physiology* 29, no. 11 (2009): 1317–28, doi:10.1093/treephys/tpp068; B. A. Black et al., "Relationships between radial growth rates and life span within North American tree species," *Ecoscience* 15, no. 3 (2008): 349–57, doi:10.2980/15-3-3149; C. Bigler and T. T. Veblen, "Increased early growth rates decrease longevities of conifers in subalpine forests," *Oikos* 118, no. 8 (2009): 1130–38, doi:10.1111/j.1600-0706.2009.17592.x.

29. *when investigators stressed the plants by removing leaves:* K. E. Rose et al., "The costs and benefits of fast living," *Ecology Letters* 12, no. 12 (2009): 1379–84, doi:10.1111/j.1461-0248.2009.01394.x.

30. *seen in laboratory studies of* Caenorhabditis elegans: W. A. Van Voorhies et al., "The longevity of *Caenorhabditis elegans* in soil," *Biology Letters* 1, no. 2 (2005): 247–49, doi:10.1098/rsbl.2004.0278; D. W. Walker et al., "Natural selection: Evolution of life span in *C. elegans*," *Nature* 405, no. 6784 (2000): 296–97.

31. *a study of senescence in long-leaved plantain:* D. A. Roach, "Environmental effects on age-dependent mortality: A test with a perennial plant species under natural and protected conditions," *Experimental Gerontology* 36, no. 4–6 (2001): 687–94.

32. *Outeniqua yellowwood: "Afrocarpus falcatus," Gymnosperm Database*, ed. C. J. Earle, accessed December 21, 2012, http://www.conifers.org/po/Afrocarpus_falcatus.php.

33. *have been estimated to be 11,700 years old:* F. C. Vasek, "Creosote bush: Long-lived clones in the Mohave desert," *American Journal of Botany* 67 (1980): 246–55.

34. *Clonal plants can reach enormous ages:* S. Arnaud-Haond et al., "Implications of extreme life span in clonal organisms: Millenary clones in meadows of the threatened seagrass *Posidonia oceanica*," *PLoS One* 7, no. 2 (2012): e30454.

35. *should not be regarded as old:* E. Clarke, "Plant individuality: A solution to the demographer's dilemma," *Biological Philosophy* (2012), doi:10.1007/s10539-012-9309-3.

36. *clones that are hundreds of years old:* E. Oinonen, "The correlation between the size of Finnish bracken (*Pteridium aquilinum* (L.) Kuhn) clones and certain periods of site history," *Acta Forestalia Fennica* 83 (1967): 1–51.

37. *clones that were up to 10,000 years old:* D. Ally et al., "Aging in a long-lived clonal tree," *PLoS Biology* 8, no. 8 (2010), doi:e100045410.1371/journal.pbio.1000454.

38. *the rapid loss of fertility in men:* S. Jones, *Y: The Descent of Men* (Little Brown, 2002), 74.

39. *there are flowering genes:* M. C. Albani and G. Coupland, "Comparative analysis of flowering in annual and perennial plants," in "Plant development," ed. M. C. P. Timmermans, *Current Topics in Developmental Biology* 91 (2010): 323–48; doi:10.1016/S0070-2153(10)91011-9; R. Amasino, "Floral induction and monocarpic versus polycarpic life histories," *Genome Biology* 10, no. 7 (2009), doi:22810.1186/gb-2009-10-7-228; S. Melzer et al., "Flowering-time genes modulate meristem determinacy and growth form in *Arabidopsis thaliana*," *Nature Genetics* 40, no. 12 (2008): 1489–92, doi:10.1038/ng.253; J. Silvertown, "A binary classification of plant life histories and some possibilities for its evolutionary application," *Evolutionary Trends in Plants* 3 (1989): 87–90; H. Thomas et al., "Annuality, perenniality and cell death," *Journal of Experimental Botany* 51, no. 352 (2000): 1781–88.

Chapter Six

1. *"E, I sing for Evolution":* Steve Knightly, "Evolution," on *Arrogance, Ignorance and Greed* (2009), by Show of Hands, Hands on Music, HMCD 29.

2. *a traditional tale of the Hausa:* T. R. Cole and M. G. Winkler, eds., *The Oxford Book of Aging* (Oxford University Press, 1994), 259.

3. *"Death is a Dialogue":* First verse of poem no. 976, in E. Dickinson, *The Complete Poems of Emily Dickinson*, ed. T. H. Johnson (Little Brown, 1960), 456.

4. *"Death be not proud": Poems of John Donne,* ed. E. K. Chambers (Lawrence & Bullen, 1896), Kindle edition.

5. *a bet with another poet:* W. Davies and R. Maud, eds., *Dylan Thomas Collected Poems 1934–1953* (Dent, 1994) (poem, 56; commentary, 208–9).

6. *"After death nothing is":* From Seneca's "Troades," trans. John Wilmot, Earl of Rochester (1647–1680), in J. Wilmot, *The Works of the Earl of Rochester* (Wordsworth Editions, 1995).

7. *nineteenth-century German biologist August Weismann:* A. Weismann, *Essays upon Heredity and Kindred Biological Problems* (Clarendon Press, 1891).

8. *British biologist Peter Medawar:* P. B. Medawar, *The Uniqueness of the Individual* (Methuen, 1957); P. B. Medawar, "Old age and natural death," *Modern Quarterly*, vol. 2 (1946): 30–56.

9. *frequency of the ε4 allele:* F. Drenos and T. B. L. Kirkwood, "Selection on alleles affecting human longevity and late-life disease: The example of apolipoprotein E," *PLoS One* 5, no. 3 (2010), doi:e1002210.1371/journal.pone.0010022.

10. *related to the immune system:* C. E. Finch, *The Biology of Human Longevity* (Academic Press, 2007).

11. *mutations that increase susceptibility to rheumatoid arthritis:* E. Corona et al., "Extreme evolutionary disparities seen in positive selection across seven complex diseases," *PLoS One* 5, no. 8 (2010), doi:e1223610.1371/journal.pone.0012236.

12. *exposed humans to many new diseases:* J. Diamond, *Guns, Germs and Steel* (Chatto & Windus, 1997).

13. *American biologist George C. Williams:* G. C. Williams, "Pleiotropy, natural selection, and the evolution of senescence," *Evolution* 11 (1957): 398–411.

14. *a Mrs. Rajo Devi gave birth at the age of 70:* Devendra Uppal, "Childless for 50 yrs, mother at 70," *Hindustan Times*, December 8, 2008, http://www.hindustantimes.com/News-Feed/haryana/Childless-for-50-yrs-mother-at-70/Article1-356574.aspx.

15. *record of reproduction back 1,200 years:* D. E. L. Promislow, "Longevity and the barren aristocrat," *Nature* 396, no. 6713 (1998): 719–20.

16. *A study of two villages in Gambia:* D. P. Shanley et al., "Testing evolutionary theories of menopause," *Proceedings of the Royal Society of London, Series B: Biological Sciences* 274, no. 1628 (2007): 2943–49, doi:10.1098/rspb.2007.1028.

17. *births and deaths in premodern Finland:* M. Lahdenpera et al., "Fitness benefits of prolonged post-reproductive life span in women," *Nature* 428, no. 6979 (2004): 178–81.

18. *survival of grandfathers:* M. Lahdenpera et al., "Selection for long life span in men: Benefits of grandfathering?," *Proceedings of the Royal Society of London, Series B: Biological Sciences* 274, no. 1624 (2007): 2437–44.

19. *experienced a fourteenfold increase in mortality:* E. A. Foster et al., "Adaptive prolonged postreproductive life span in killer whales," *Science* 337, no. 6100 (2012): 1313, doi:10.1126/science.1224198.

20. *Without such a close-knit family structure:* R. A. Johnstone and M. A. Cant, "The evolution of menopause in cetaceans and humans: The role of demography," *Proceedings of the Royal Society of London, Series B: Biological Sciences* 277, no. 1701 (2010): 3765–71, doi:10.1098/rspb.2010.0988.

21. *women are the more robust sex:* S. N. Austad, "Why women live longer than men: Sex differences in longevity," *Gender Medicine* 3, no. 2 (2006): 79–92.

22. *viruses:* M. De Paepe and F. Taddei, "Viruses' life history: Towards a mechanistic basis of a trade-off between survival and reproduction among phages," *PLoS Biology* 4, no. 7 (2006): 1248–56, doi:e19310.1371/journal.pbio.0040193.

23. *every species at which anyone has ever looked:* W. A. Van Voorhies et al.,

"Do longevity mutants always show trade-offs?," *Experimental Gerontology* 41, no. 10 (2006): 1055–58, doi:10.1016/j.exger.2006.05.006.

24. *Arnold Schoenberg summarized his art:* D. Zanette, "Playing by numbers," *Nature* 453 (June 19, 2008): 988–89.

25. *longer-lived mutants disappeared:* N. L. Jenkins et al., "Fitness cost of extended life span in *Caenorhabditis elegans*," *Proceedings of the Royal Society of London, Series B: Biological Sciences* 271, no. 1556 (2004): 2523–26, doi:10.1098/rspb.2004.2897.

26. *Another* C. elegans *longevity gene, called* clk-1: J. Chen et al., "A demographic analysis of the fitness cost of extended longevity in *Caenorhabditis elegans*," *Journals of Gerontology, Series A, Biological Sciences and Medical Sciences* 62, no. 2 (2007): 126–35.

27. C. elegans *that were fed with resveratrol:* J. Gruber et al., "Evidence for a trade-off between survival and fitness caused by resveratrol treatment of *Caenorhabditis elegans*," in *Biogerontology: Mechanisms and Interventions*, ed. S. I. S. Rattan and S. Akman, Annals of the New York Academy of Sciences, 1100 (New York Academy of Sciences, 2007), 530–42.

28. *fruit flies lacking ovaries lived significantly longer:* J. Maynard Smith, "The effects of temperature and of egg-laying on the longevity of *Drosophila subobscura*," *Journal of Experimental Biology* 35 (1958): 832–42.

29. *reproductive cells generate chemical signals:* T. Flatt, "Survival costs of reproduction in *Drosophila*," *Experimental Gerontology* 46, no. 5 (2011): 369–75, doi:10.1016/j.exger.2010.10.008.

30. *in the natural environment of the soil:* W. A. Van Voorhies et al., "The longevity of *Caenorhabditis elegans* in soil," *Biology Letters* 1, no. 2 (2005): 247–49, doi:10.1098/rsbl.2004.0278.

31. *animals in zoos are cosseted like royalty:* R. E. Ricklefs and C. D. Cadena, "Lifespan is unrelated to investment in reproduction in populations of mammals and birds in captivity," *Ecology Letters* 10, no. 10 (2007): 867–72; R. E. Ricklefs and C. D. Cadena, "Rejoinder to Ricklefs and Cadena (2007): Response to Mace and Pelletier," *Ecology Letters* 10, no. 10 (2007): 874–75, doi:10.1111/j.1461-0248.2007.01103.x.

Chapter Seven

1. *"I shall live in my fame":* Ovid, *Metamorphoses* (Penguin, 2004).

2. *literature is suffused with the influence of Ovid's* Metamorphoses: S. A. Brown, *Ovid: Myth and Metamorphosis* (Bristol Classical Press, 2005).

3. *Among the works of this prolific author:* George Frideric Handel, *Semele*, performed by Monteverde Choir & English Baroque soloists, conducted by John Elliot Gardiner, sleeve notes, released February 3, 1993, Erato 2292-45982-2, 1993.

4. *Eels make a one-way trip:* T. Fort, *The Book of Eels* (Harper Collins, 2002).

5. *Many squid and octopus species are semelparous:* F. Rocha et al., "A review of reproductive strategies in cephalopods," *Biological Reviews* 76, no. 3 (2001): 291–304; L. C. Hendrickson and D. R. Hart, "An age-based cohort model for estimating the spawning mortality of semelparous cephalopods with an application to per-recruit calculations for the northern shortfin squid, *Illex illecebrosus*," *Cephalopod Stock Assessment Workshop* (2004): 4–13, doi:10.1016/j.fishres.2005.12.005.

6. *Some snakes are semelparous:* R. Shine, "Reproductive strategies in snakes," *Proceedings of the Royal Society of London, Series B: Biological Sciences* 270, no. 1519 (2003): 995–1004, doi:10.1098/rspb.2002.2307; K. B. Karsten et al., "A unique life history among tetrapods: An annual chameleon living mostly as an egg," *Proceedings of the National Academy of Sciences of the United States of America* 105, no. 26 (2008): 8980–84, doi:10.1073/pnas.0802468105.

7. *American biologist Lamont C. Cole:* L. C. Cole, "The population consequences of life history phenomena," *Quarterly Review of Biology* 29, no. 2 (1954): 103–37, doi:10.1086/400074.

8. *The rule is that for a repeat breeder:* M. Bulmer, *Theoretical Evolutionary Ecology* (Sinauer Associates, 1994).

9. *and animals now reproduce precociously:* M. E. Jones et al., "Life-history change in disease-ravaged Tasmanian devil populations," *Proceedings of the National Academy of Sciences of the United States of America* 105, no. 29 (2008): 10023–27, doi:10.1073/pnas.0711236105.

10. *in the brown antechinus:* C. E. Holleley et al., "Size breeds success: Multiple paternity, multivariate selection and male semelparity in a small marsupial, *Antechinus stuartii*," *Molecular Ecology* 15, no. 11 (2006): 3439–48, doi:10.1111/j.1365-294X.2006.03001.x.

11. *Males' physiology overdoses on testosterone:* R. Naylor et al., "Boom and bust: A review of the physiology of the marsupial genus *Antechinus*," *Journal of Comparative Physiology B: Biochemical, Systemic, and Environmental Physiology* 178, no. 5 (2008): 545–62, doi:10.1007/s00360-007-0250-8; M. Wolkewitz et al., "Is 27 really a dangerous age for famous musicians? Retrospective cohort study," *British Medical Journal* 343 (2011), doi:10.1136/bmj.d7799.

12. *favors multiple mating:* Wolkewitz et al., "Is 27 really a dangerous age?;

K. Kraaijeveld et al., "Does female mortality drive male semelparity in dasyurid marsupials?," *Proceedings of the Royal Society of London, Series B: Biological Sciences* 270 (2003): S251–S253.

13. *some of those males were not semelparous:* K. M. Wolfe et al., "Post-mating survival in a small marsupial is associated with nutrient inputs from seabirds," *Ecology* 85, no. 6 (2004): 1740–46.

14. *The capelin is a marine fish:* J. S. Christiansen et al., "Facultative semelparity in capelin *Mallotus villosus* (Osmeridae): An experimental test of a life history phenomenon in a sub-arctic fish," *Journal of Experimental Marine Biology and Ecology* 360, no. 1 (2008): 47–55, doi:10.1016/j.jembe.2008.04.003.

15. *insects and spiders:* D. W. Tallamy and W. P. Brown, "Semelparity and the evolution of maternal care in insects," *Animal Behaviour* 57 (1999): 727–30.

16. *Females of the crab spider:* K. Futami and S. Akimoto, "Facultative second oviposition as an adaptation to egg loss in a semelparous crab spider," *Ethology* 111, no. 12 (2005): 1126–38.

17. *Japanese hump earwig:* S. Suzuki et al., "Matriphagy in the hump earwig, *Anechura harmandi* (Dermaptera: Forficulidae), increases the survival rates of the offspring," *Journal of Ethology* 23, no. 2 (2005): 211–13, doi:10.1007/s10164-005-0145-7.

18. *Fish then swim to the shore:* I. A. Fleming and M. R. Gross, "Evolution of adult female life history and morphology in a Pacific salmon (coho: *Oncorhynchus kisutch*)," *Evolution* 43, no. 1 (1989): 141–57.

19. *Salmon use Earth's magnetic field:* K. J. Lohmann et al., "Geomagnetic imprinting: A unifying hypothesis of long-distance natal homing in salmon and sea turtles," *Proceedings of the National Academy of Sciences of the United States of America* 105, no. 49 (2008): 19096–101, doi:10.1073/pnas.0801859105; H. Bandoh et al., "Olfactory responses to natal stream water in sockeye salmon by BOLD fMRI," *PLoS One* 6, no. 1 (2011), doi:10.1371/journal.pone.0016051.

20. *transfer of nutrients from salmon:* M. D. Hocking and J. D. Reynolds, "Impacts of salmon on riparian plant diversity," *Science* 331, no. 6024 (2011): 1609–12, doi:10.1126/science.1201079.

21. *Predation has such a strong effect:* S. M. Carlson et al., "Predation by bears drives senescence in natural populations of salmon," *PLoS One* 2, no. 12 (2007), doi:10.1371/journal.pone.0001286.

22. *a greater weight of eggs:* B. J. Crespi and R. Teo, "Comparative phylogenetic analysis of the evolution of semelparity and life history in salmonid fishes," *Evolution* 56, no. 5 (2002): 1008–20.

23. *the Atlantic species is a repeat breeder:* I. A. Fleming, "Reproductive strategies of Atlantic salmon: Ecology and evolution," *Reviews in Fish Biology*

and Fisheries 6, no. 4 (1996): 379–416, doi:10.1007/bf00164323; C. Garcia de Leaniz et al., "A critical review of adaptive genetic variation in Atlantic salmon: Implications for conservation," *Biological Reviews* 82, no. 2 (2007): 173–211, doi:10.1111/j.1469-185X.2006.00004.x.

24. *fewer than one in ten:* Fleming, "Reproductive strategies of Atlantic salmon."

25. *which gives the jacks a relative advantage:* M. R. Gross, "Disruptive selection for alternative life histories in salmon," *Nature* 313 (1985): 47–48; Y. Tanaka et al., "Breeding games and dimorphism in male salmon," *Animal Behaviour* 77, no. 6 (2009): 1409–13, doi:10.1016/j.anbehav.2009.01.039.

26. *jacks are delayed in migrating:* M. Buoro et al., "Investigating evolutionary trade-offs in wild populations of Atlantic salmon (*Salmo salar*): Incorporating detection probabilities and individual heterogeneity," *Evolution* 64, no. 9 (2010): 2629–42, doi:10.1111/j.1558-5646.2010.01029.x.

27. *bamboos achieve flowering synchrony:* D. H. Janzen, "Why bamboos wait so long to flower," *Annual Review of Ecology and Systematics* 7 (1976): 347–91.

28. *Giant pandas feed exclusively on the leaves of semelparous bamboos:* J. Carter et al., "Giant panda (*Ailuropoda melanoleuca*) population dynamics and bamboo (subfamily Bambusoideae) life history: A structured population approach to examining carrying capacity when the prey are semelparous," *Ecological Modelling* 123, no. 2–3 (1999): 207–23; K. G. Johnson et al., "Responses of giant pandas to a bamboo die-off," *National Geographic Research* 4 (1988): 161–77.

29. *decaying bodies create a pulse of nitrogen:* L. H. Yang, "Periodical cicadas as resource pulses in North American forests," *Science* 306, no. 5701 (2004): 1565–67.

30. *The century plant* Agave americana: M. Rocha et al., "Reproductive ecology of five sympatric *Agave littaea* (Agavaceae) species in Central Mexico," *American Journal of Botany* 92, no. 8 (2005): 1330–41.

Chapter Eight

1. *"Every night I'm in a different town":* Venom, "Live Like an Angel," on *Welcome to Hell* (1981), accessed September 13, 2012, http://lyrics.rockmagic.net/lyrics/venom/welcome_to_hell_1981.html#s05.

2. *many of the kind die at 27: Wikipedia*, s.v. "The 27 Club," accessed September 13, 2012, http://en.wikipedia.org/wiki/27_Club.

3. *alcohol poisoning:* "Winehouse died from alcohol poisoning after going on drinking binge": *Guardian*, October 27, 2011, 5.

4. *propensity to die at age 27:* M. Wolkewitz et al., "Is 27 really a dangerous age for famous musicians? Retrospective cohort study," *British Medical Journal* 343 (2011), doi:10.1136/bmj.d7799.

5. *shrew burns energy at twenty-five times the rate of a rock star:* D. W. MacDonald, ed., *The New Encyclopedia of Mammals* (Oxford University Press, 2001).

6. *more than 600 beats per minute:* J. T. Bonner, *Why Size Matters* (Princeton University Press, 2006), 117.

7. *he didn't always get the numbers:* I. L. Goldman, "Raymond Pearl, smoking and longevity," *Genetics* 162, no. 3 (2002): 997–1001.

8. *Although the patients died:* R. Pearl, "Cancer and tuberculosis," *American Journal of Hygiene* 9, no. 1 (1929): 97–159; R. Pearl et al., "Experimental treatment of cancer with tuberculin," *Lancet* 1 (1929): 1078–80.

9. *Pearl pursued a mathematical solution:* H. S. Jennings, "Biographical memoir of Raymond Pearl, 1879–1940," *National Academy of the United States of America Biographical Memoirs* 22, no. 14 (1942): 294–347.

10. *destroyed by a lab fire:* R. Pearl, "An appeal," *Science (New York, NY)* 50, no. 1301 (1919): 524–25, doi:10.1126/science.50.1301.524-a.

11. *his French horn reportedly "blew up":* S. E. Kingsland, "Raymond Pearl: On the frontier in the 1920s—Raymond Pearl Memorial Lecture (1983)," *Human Biology* 56, no. 1 (1984): 1–18.

12. *beer was brewed clandestinely:* S. Mayfield, *The Constant Circle: H. L. Mencken and His Friends* (Delacorte Press, 1968).

13. *effects on the growth of seedlings:* R. Pearl and A. Allen, "The influence of alcohol upon the growth of seedlings," *Journal of General Physiology* 8, no. 3 (1926): 215–31, doi:10.1085/jgp.8.3.215.

14. *modest imbibing can lengthen life:* R. Lakshman et al., "Is alcohol beneficial or harmful for cardioprotection?," *Genes and Nutrition* 5, no. 2 (2010): 111–20, doi:10.1007/s12263-009-0161-2.

15. *moderate smoking was harmful to longevity:* R. Pearl, "Studies on human longevity VII. Tobacco smoking and longevity," *Science* 87 (1938): 216–17.

16. *his 1926 book* Alcohol and Longevity: R. Pearl, *Alcohol and Longevity* (Alfred Knopf, 1926).

17. *novel published in 1925 by Sinclair Lewis:* H. S. Lewis, *Arrowsmith* (New American Library, 1925), 387.

18. *In his book* The Rate of Living: R. Pearl, *The Rate of Living, Being an Account of Some Experimental Studies on the Biology of Life Duration* (Alfred Knopf, 1928).

19. *his lecture series "The Biology of Death"*: R. Pearl, *The Biology of Death* (J. B. Lippincott, 1922).

20. *a popular article in the* Baltimore Sun: S. N. Austad, *Why We Age* (Wiley, 1997), 76.

21. *cooler water fleas lived longer:* J. W. MacArthur and W. H. T. Baillie, "Metabolic activity and duration of life II. Metabolic rates and their relation to longevity in *Daphnia magna*," *Journal of Experimental Zoology* 53, no. 2 (1929): 243–68, doi:10.1002/jez.1400530206.

22. *answer finally just popped into his head:* K. Kitani and G. O. Ivy, "I thought, thought, thought for four months in vain and suddenly the idea came"—An interview with Denham and Helen Harman," *Biogerontology* 4, no. 6 (2003): 401–12, doi:10.1023/b:bgen.0000006561.15498.68.

23. *for nearly a decade after he published:* D. Harman, "Aging: A theory based on free-radical and radiation chemistry," *Journal of Gerontology* 11, no. 3 (1956): 298–300.

24. *whether this damage is the most important cause of aging:* A. A. Freitas and J. P. de Magalhães, "A review and appraisal of the DNA damage theory of ageing," *Mutation Research—Reviews in Mutation Research* 728, no. 1–2 (2011): 12–22, doi:10.1016/j.mrrev.2011.05.001.

25. *"That on the ashes of his youth doth lie":* William Shakespeare, Sonnet no. 73, in *The Complete Works of William Shakespeare*, Royal Shakespeare Company Edition, ed. J. Bate and E. Rasmussen (Macmillan, 2006).

26. *uncovering its living mechanism:* K. B. Beckman and B. N. Ames, "The free radical theory of aging matures," *Physiological Reviews* 78, no. 2 (1998): 547–81.

27. *conserve energy by hibernation:* S. N. Austad and K. E. Fischer, "Mammalian aging, metabolism, and ecology: Evidence from the bats and marsupials," *Journals of Gerontology, Biological Sciences* 46, no. 2 (1991): B47–B53.

28. *Birds show an even more deviant pattern than bats:* D. J. Holmes et al., "Comparative biology of aging in birds: An update," *Experimental Gerontology* 36, no. 4–6 (2001): 869–83, doi:10.1016/s0531-5565(00)00247-3.

29. *a database called AnAge:* AnAge: The Animal Ageing and Longevity Database, accessed December 30, 2011, http://genomics.senescence.info/species/.

30. *no correlation between longevity and metabolic rate:* J. P. de Magalhães et al., "An analysis of the relationship between metabolism, developmental schedules, and longevity using phylogenetic independent contrasts," *Journals of Gerontology, Series A, Biological Sciences and Medical Sciences* 62, no. 2 (2007): 149–60.

31. *longer life in subterranean mammals:* R. M. Sibly and J. H. Brown, "Effects of body size and lifestyle on evolution of mammal life histories," *Proceedings of the National Academy of Sciences of the United States of America* 104, no. 45 (2007): 17707–12, doi:10.1073/pnas.0707725104.

32. *chemical defenses that make an animal unpalatable:* M. A. Blanco and P. W. Sherman, "Maximum longevities of chemically protected and non-protected fishes, reptiles, and amphibians support evolutionary hypotheses of aging," *Mechanisms of Ageing and Development* 126, no. 6–7 (2005): 794–803, doi:10.1016/j.mad.2005.02.006.

33. *hibernation:* C. Turbill et al., "Hibernation is associated with increased survival and the evolution of slow life histories among mammals," *Proceedings of the Royal Society of London, Series B: Biological Sciences* 278, no. 1723 (2011): 3355–63, doi:10.1098/rspb.2011.0190.

34. *living in trees:* M. R. Shattuck and S. A. Williams, "Arboreality has allowed for the evolution of increased longevity in mammals," *Proceedings of the National Academy of Sciences of the United States of America* 107, no. 10 (2010): 4635–39, doi:10.1073/pnas.0911439107.

35. *body armor:* J. W. Gibbons, "Why do turtles live so long?," *BioScience* 37, no. 4 (1987): 262–69, doi:10.2307/1310589.

36. *George C. Williams predicted exactly such a pattern:* G. C. Williams, "Pleiotropy, natural selection, and the evolution of senescence," *Evolution* 11 (1957): 398–411.

37. *available data for birds and mammals:* R. E. Ricklefs, "Evolutionary theories of aging: Confirmation of a fundamental prediction, with implications for the genetic basis and evolution of life span," *American Naturalist* 152 (1998): 24–44.

38. *same generation times senesce at the same rate:* O. R. Jones et al., "Senescence rates are determined by ranking on the fast-slow life-history continuum," *Ecology Letters* 11, no. 7 (2008): 664–73, doi:10.1111/j.1461-0248.2008.01187.x.

39. *really put to an unequivocal test:* S. C. Stearns et al., "Experimental evolution of aging, growth, and reproduction in fruitflies," *Proceedings of the National Academy of Sciences of the United States of America* 97, no. 7 (2000): 3309–13.

40. *earlier reproduction in flies:* T. Flatt, "Survival costs of reproduction in *Drosophila*," *Experimental Gerontology* 46, no. 5 (2011): 369–75, doi:10.1016/j.exger.2010.10.008.

41. *classified as a single species:* M. O. Winfield et al., "A brief evolutionary excursion comes to an end: The genetic relationship of British species of *Gen-*

tianella sect. *Gentianella* (Gentianaceae)," *Plant Systematics and Evolution* 237, no. 3–4 (2003): 137–51, doi:10.1007/s00606-002-0248-3.

42. *"They were falling apart":* interview with Steven N. Austad, State of Tomorrow (University of Texas Foundation), accessed January 7, 2012, http://www.stateoftomorrow.com/stories/transcripts/AustadInterviewTranscript.pdf.

43. *wandered around during the day:* S. N. Austad, *Why We Age* (Wiley, 1997), 114.

44. *rate of aging was about half:* S. N. Austad, "Retarded senescence in an insular population of Virginia opossums (*Didelphis virginiana*)," *Journal of Zoology* 229 (1993): 695–708.

45. *Primates are tree dwellers:* M. R. Shattuck and S. A. Williams, "Arboreality has allowed for the evolution of increased longevity in mammals," *Proceedings of the National Academy of Sciences of the United States of America* 107, no. 10 (2010): 4635–39, doi:10.1073/pnas.0911439107.

46. *species with bigger brains live longer:* C. Gonzalez-Lagos et al., "Large-brained mammals live longer," *Journal of Evolutionary Biology* 23, no. 5 (2010): 1064–74, doi:10.1111/j.1420-9101.2010.01976.x.

Chapter Nine

1. *Robert Heinlein's science fiction novel:* R. A. Heinlein, *Methuselah's Children* (New English Library, 1980), originally published 1941.

2. *life span has advanced by nearly 15 minutes per hour:* J. Oeppen and J. W. Vaupel, "Demography: Broken limits to life expectancy," *Science* 296, no. 5570 (2002): 1029–31.

3. *failed to show any clear benefits:* D. Giustarini et al., "Oxidative stress and human diseases: Origin, link, measurement, mechanisms, and biomarkers," *Critical Reviews in Clinical Laboratory Sciences* 46, no. 5–6 (2009): 241–81, doi:10.3109/10408360903142326.

4. *oxygen free radicals are not just dangerous by-products:* J. P. de Magalhães and G. Church, "Cells discover fire: Employing reactive oxygen species in development and consequences for aging," *Experimental Gerontology* 41, no. 1 (2006): 1–10, doi:10.1016/j.exger.2005.09.002.

5. *the ocean quahog:* Z. Ungvari et al., "Extreme longevity is associated with increased resistance to oxidative stress in *Arctica islandica*, the longest-living non-colonial animal," *Journals of Gerontology, Series A, Biological Sciences and Medical Sciences* 66, no. 7 (2011): 741–50, doi:10.1093/gerona/glr044.

6. *tiny cave-dwelling olm salamander:* J. Issartel et al., "High anoxia tolerance in the subterranean salamander *Proteus anguinus* without oxidative stress

nor activation of antioxidant defenses during reoxygenation," *Journal of Comparative Physiology B: Biochemical, Systemic, and Environmental Physiology* 179, no. 4 (2009): 543–51, doi:10.1007/s00360-008-0338-9.

7. *tolerate these levels of stress:* K. N. Lewis et al., "Stress resistance in the naked mole-rat: The bare essentials: A mini-review," *Gerontology* 58, no. 5 (2012): 453–62.

8. *no effect on how long the animals live:* J. R. Speakman and C. Selman, "The free-radical damage theory: Accumulating evidence against a simple link of oxidative stress to ageing and life span," *Bioessays* 33, no. 4 (2011): 255–59, doi:10.1002/bies.201000132.

9. *signals which males are best fortified:* T. von Schantz et al., "Good genes, oxidative stress and condition-dependent sexual signals," *Proceedings of the Royal Society of London, Series B: Biological Sciences* 266, no. 1414 (1999): 1–12, doi:10.1098/rspb.1999.0597.

10. *males that females preferred:* C. R. Freeman-Gallant et al., "Oxidative damage to DNA related to survivorship and carotenoid-based sexual ornamentation in the common yellowthroat," *Biology Letters* 7, no. 3 (2011): 429–32, doi:10.1098/rsbl.2010.1186.

11. *survived significantly longer:* N. Saino et al., "Antioxidant defenses predict long-term survival in a passerine bird," *PLoS One* 6, no. 5 (2011), doi: e1959310.1371/journal.pone.0019593.

12. *reproductive success is correlated with carotenoid concentration:* R. J. Safran et al., "Positive carotenoid balance correlates with greater reproductive performance in a wild bird," *PLoS One* 5, no. 2 (2010), doi:e942010.1371/journal .pone.0009420.

13. *"senescence should always be a generalized deterioration":* G. C. Williams, "Pleiotropy, natural selection, and the evolution of senescence," *Evolution* 11 (1957): 398–411.

14. *everything except the germ line senesces:* R. Holliday and S. I. S. Rattan, "Longevity mutants do not establish any 'new science' of ageing," *Biogerontology* 11, no. 4 (2010): 507–11, doi:10.1007/s10522-010-9288-1.

15. *Aubrey de Grey, a maverick from Cambridge:* J. Weiner, *Long for This World: The Strange Science of Immortality* (Ecco, 2010).

16. *"Strategies for Engineered Negligible Senescence"*: A. de Grey, "Defeat of aging: Utopia or foreseeable scientific reality," in *Future of Life and the Future of Our Civilization*, ed. V. Burdyuzha (Springer 2006), 277–90.

17. *it comprises ten separate diseases:* C. Curtis et al., "The genomic and transcriptomic architecture of 2,000 breast tumours reveals novel subgroups," *Nature* 486, no. 7403 (2012), 346–52, doi:10.1038/nature10983.

18. *discovered by Leonard Hayflick:* L. Hayflick and P. S. Moorhead, "Serial cultivation of human diploid cell strains," *Experimental Cell Research* 25, no. 3 (1961): 585–621, doi:10.1016/0014-4827(61)90192-6.

19. *it seemed like an obvious cause of aging:* J. W. Shay and W. E. Wright, "Hayflick, his limit, and cellular ageing," *Nature Reviews Molecular Cell Biology* 1, no. 1 (2000): 72–76.

20. *a structure involved with the replication of DNA:* E. H. Blackburn et al., "Telomeres and telomerase: The path from maize, *Tetrahymena* and yeast to human cancer and aging," *Nature Medicine* 12, no. 10 (2006): 1133–38.

21. *between six and nine feet long:* S. Chen, "Length of a human DNA molecule," in *The Physics Factbook*, ed. Glenn Elert, accessed January 25, 2012, http://hypertextbook.com/facts/1998/StevenChen.shtml.

22. *replicative senescence limits life span:* L. Hayflick, "Human cells and aging," *Scientific American* 218, no. 3 (1968): 32–37.

23. *mouse cells can replicate indefinitely in the lab:* K. A. Mather et al., "Is telomere length a biomarker of aging? A review," *Journals of Gerontology, Series A, Biological Sciences and Medical Sciences* 66, no. 2 (2011): 202–13, doi:10.1093/gerona/glq180.

24. *all cancer cells produce telomerase:* J. W. Shay and W. E. Wright, "Role of telomeres and telomerase in cancer," *Seminars in Cancer Biology* 21, no. 6 (2011): 349–53, doi:10.1016/j.semcancer.2011.10.001.

25. *telomerase activity in fifteen different rodent species:* A. Seluanov et al., "Telomerase activity coevolves with body mass not life span," *Aging Cell* 6, no. 1 (2007): 45–52, doi:10.1111/j.1474-9726.2006.00262.x.

26. *critical size at which telomerase becomes a costly cancer risk:* N. M. V. Gomes et al., "Comparative biology of mammalian telomeres: Hypotheses on ancestral states and the roles of telomeres in longevity determination," *Aging Cell* 10, no. 5 (2011): 761–68, doi:10.1111/j.1474-9726.2011.00718.x.

27. *individuals with longer telomeres:* P. Bize et al., "Telomere dynamics rather than age predict life expectancy in the wild," *Proceedings of the Royal Society of London, Series B: Biological Sciences* 276, no. 1662 (2009): 1679–83, doi:10.1098/rspb.2008.1817; C. M. Vleck et al., "Evolutionary ecology of senescence: A case study using tree swallows, *Tachycineta bicolor*," *Journal of Ornithology* 152 (2011): 203–11, doi:10.1007/s10336-010-0629-2; H. M. Salomons et al., "Telomere shortening and survival in free-living corvids," *Proceedings of the Royal Society of London, Series B: Biological Sciences* 276, no. 1670 (2009): 3157–65, doi:10.1098/rspb.2009.0517; C. G. Foote et al., "Individual state and survival prospects: Age, sex, and telomere length in a long-lived seabird," *Behavioral Ecology* 22, no. 1 (2011): 156–61, doi:10.1093/beheco/arq178.

28. *mortality and telomere length:* R. M. Cawthon et al., "Association between telomere length in blood and mortality in people aged 60 years or older," *Lancet* 361, no. 9355 (2003): 393–95.

29. *a review of those studies conducted in 2011:* Mather et al., "Is telomere length a biomarker of aging?"

30. *how good you look for your age:* D. A. Gunn et al., "Perceived age as a biomarker of ageing: A clinical methodology," *Biogerontology* 9, no. 5 (2008): 357–64, doi:10.1007/s10522-008-9141-y.

31. *removing senescent cells:* D. J. Baker et al., "Clearance of p16Ink4a-positive senescent cells delays ageing-associated disorders," *Nature* 479, no. 7372 (2011): 232–36.

32. *induced senescent human cells to divide:* L. Lapasset et al., "Rejuvenating senescent and centenarian human cells by reprogramming through the pluripotent state," *Genes & Development* 25, no. 21 (2011): 2248–53, doi:10.1101/gad.173922.111.

33. *inequality of incomes:* R. Wilkinson and K. Pickett, *The Spirit Level: Why More Equal Societies Almost Always Do Better* (Penguin Books, 2010).

34. *the gap between rich and poor is large:* Wilkinson and Pickett, *The Spirit Level.*

INDEX